A PHILOSOPF

Why do gardens matter so much
the intriguing question to which David Cooper seeks an answer in this
book. Given the enthusiasm for gardens in human civilization ancient
and modern, Eastern and Western, it is surprising that the question has
been so long neglected by modern philosophy. Now at last there is a
philosophy of gardens. Not only is this a fascinating subject in its own
right, it also provides a reminder that the subject-matter of aesthetics is
broader than the fine arts; that ethics is not just about moral issues but
about 'the good life'; and that environmental philosophy should not
focus only on 'wilderness' to the exclusion of the humanly shaped
environment.

David Cooper identifies garden appreciation as a special human
phenomenon distinct from both from the appreciation of art and
the appreciation of nature. He explores the importance of various
'garden-practices' and shows how not only gardening itself, but
activities to which the garden especially lends itself, including
social and meditative activities, contribute to the good life. And he
distinguishes the many kinds of meanings that gardens may have,
from representation of nature to emotional expression, from historical significance to symbolization of a spiritual relationship to
the world. Building on the familiar observation that, among
human beings' creations, the garden is peculiarly dependent on
the co-operation of nature, Cooper argues that the garden matters
as an epiphany of an intimate co-dependence between human
creative activity in the world and the 'mystery' that allows there to
be a world for them at all.

A Philosophy of Gardens will open up this subject to students and
scholars of aesthetics, ethics, and cultural and environmental studies,
and to anyone with a reflective interest in things horticultural.

David E. Cooper is Professor of Philosophy at Durham University

A Philosophy of
Gardens

David E. Cooper

CLARENDON PRESS · OXFORD

OXFORD

UNIVERSITY PRESS

Great Clarendon Street, Oxford OX2 6DP

Oxford University Press is a department of the University of Oxford.
It furthers the University's objective of excellence in research, scholarship,
and education by publishing worldwide in

Oxford New York

Auckland Cape Town Dar es Salaam Hong Kong Karachi
Kuala Lumpur Madrid Melbourne Mexico City Nairobi
New Delhi Shanghai Taipei Toronto

With offices in

Argentina Austria Brazil Chile Czech Republic France Greece
Guatemala Hungary Italy Japan Poland Portugal Singapore
South Korea Switzerland Thailand Turkey Ukraine Vietnam

Oxford is a registered trade mark of Oxford University Press
in the UK and in certain other countries

Published in the United States
by Oxford University Press Inc., New York

British Library Cataloguing in Publication Data

Data available

Library of Congress Cataloging in Publication Data
Cooper, David Edward.
A philosophy of gardens / David E. Cooper.
p. cm.
Includes bibliographical references (p.) and index.
1. Gardening—Philosophy. I. Title.
SB454.3.P34C66 2006   712.01—dc22   2005025712

Typeset by Laserwords Private Limited, Chennai, India
Printed in Great Britain
on acid-free paper by
Biddles Ltd., King's Lynn, Norfolk

ISBN 978–0–19–929034–5(Hbk.)   978–0–19–923888–0(Pbk.)

1 3 5 7 9 10 8 6 4 2

ACKNOWLEDGEMENTS

This book emerged out of a paper given at the 2002 annual conference of the British Society of Aesthetics (see Cooper 2003*a*), and my first thanks are to the several hearers or readers of that paper who encouraged me to work up the thoughts it expressed into a book. They will find, with or without approval, that some of the ideas developed in the book have no obvious precedent in the paper. I also thank my colleagues Dr A. J. Hamilton and Dr Simon P. James for helpful comments on various chapters, and the two anonymous publisher's readers for their remarks, at once enthusiastic and corrective, on the typescript. I am grateful, as well, to several people at Oxford University Press for helping to see the book through from conception to completion: the philosophy editors, Peter Momtchiloff and Rupert Cousens, and the Humanities Team Manager, Rebecca Bryant. Finally, I thank Jean van Altena for her impeccable copy-editing, and Bret Wallach and greatmirror.com for permission to reproduce Professor Wallach's photo of the Daitokuji garden in Kyoto.

DAVID E. COOPER

CONTENTS

Any one can create a pretty little bamboo garden in the world. But I doubt that the gardener would succeed in incorporating the world in his bamboo garden.

Hermann Hesse, The Glass Bead Game

LIST OF ILLUSTRATIONS

'A Fundamental Question'

The volume of philosophical writings about gardens in recent years is modest—so modest that their authors typically begin, as I am now doing, by remarking on the relative neglect of the garden by modern philosophy. This neglect partly explains why the title of my book starts with the indefinite article. In the absence of a substantial literature, there simply does not exist, within the philosophical community, a shared perception of a 'discipline'—*the* philosophy of gardens—replete with well-defined 'problems', 'methods', and 'research programmes'. One may speak of, and write about, the philosophy of mind—or of science or of art—but not, in an analogous way, of the philosophy of gardens. People who buy a book called *The Philosophy of Mind* expect from it an introduction to a discipline. *The Philosophy of Gardens* would be a fraudulent title: for here there is no discipline as yet to be introduced to.

In the next section, I say something about the extent of, and reasons for, this relative neglect of gardens. For the moment, and in order to help locate my main theme, I remark only that this neglect is prima facie surprising. For one thing, gardens surely invite many questions of the kind posed by contemporary philosophers of art: conceptual ones ('What *is* a garden?'); ontological ones ('Is a garden simply a complex physical object?'); normative ones ('What makes a garden successful or "great"?')—and so on. Replace 'garden' by 'artwork' in those questions, and they become very familiar ones indeed. Perhaps, though, this is one reason *for* the relative philosophical neglect of gardens. The questions are *too* close to familiar ones to inspire fresh philosophical attention. By all means, if you will, illustrate the conceptual and other problems that preoccupy modern philosophers of art with examples of gardens instead of paintings or poems—but few, if any, novel issues are thereby raised. And this is also why, in the present book, such questions and issues, while not ignored, are not the main business.

To identify what that business is, it is helpful to mention a further reason, less easily disposed of, why the relative neglect of gardens is surprising. This is the striking contrast between such neglect and the palpable importance of, and enthusiasm for, gardens in modern life. Two American authors describe gardening as their country's 'national art form', engaged in by 78 per cent of its population of a weekend (Francis and Hester 1990: 8). In the UK, observes Sir Roy Strong, an 'extraordinary Renaissance has occurred': 'gardening has become one of this island's miracles...bringing us together'—in ways that the church and monarchy have lost the power to do—'with one single shared activity' (Strong 2000a: 209, 213). Less edifyingly put, gardening is, according to one journalist, 'the new rock'n roll...very social and very, very fashionable'. Such remarks are unlikely to be challenged by those of us—that is, all of us who are old enough—who have witnessed the mushrooming over recent decades of garden centres, TV garden programmes, gardening books, gardens opened to the public, and other testaments to this 'extraordinary Renaissance'.

Of course, one hardly expects philosophy to occupy itself with everything that happens to have become 'very, very fashionable'. What is novel about the recent enthusiasm for gardens, however, is only its demotic character. The designing, making, and appreciation of gardens—and the comportment of lives within and in relation to gardens—have been of importance to men and women since the days of the ancient empires of Persia and China. In neglecting the garden, philosophy is therefore ignoring not merely a current fashion, but activities and experiences of abiding human significance. And that, surely, is puzzling.

'Why garden?', asks an authority on Japanese garden design, adding that this is 'a fundamental question yet one that is hardly ever addressed' (Keane 1996: 118). Suitably construed, this is my fundamental question too. What explains the immense significance that human beings locate in the making and experiencing of gardens? The question, as I treat it, is large as well as fundamental—though not quite so large as might be imagined: for it is not with every dimension of the significance of the garden that I shall be concerned. With pardonable exaggeration, a historian of landscape art claims that 'one could write an illuminating . . . history of a nation's cultural development by examining its changing conception of the garden's scope, design and function' (Andrews 1999: 53). But I shan't, except in passing, be concerned with the historical significance of the garden— with what different conceptions of it show us about the cultures in which they arose. Nor is the significance that concerns me of the kind recently addressed by a number of anthropologists and biologists, who speculate, for example, on how gardens may be indicative of a human need for 'the dominion of nature', or be reminiscent of the natural landscapes that were of survival value to our primitive ancestors (a useful blend of open vista and shelter, of 'prospect' and 'refuge' (see Appleton 1975)).

The kind of significance that concerns me—the kind germane to the 'fundamental question' of 'Why garden?'—is the significance gardens have *for* people: for Sir Francis Bacon, say, when judging the garden to be 'the purest of human pleasures' (Bacon 1902: 127), or for the broadcaster Alan Titchmarsh when remarking that gardening is, apart from having children, 'the most rewarding thing in life', or for Roy Strong when replying 'To have made a garden' to the question 'What is the most important thing I've done with my life?' (Strong 2000*a*: 213). Significance, so construed, must be *available* to people, something they may recognize gardens as having for them, and something, therefore, that is a reason *for them* to make, appreciate, and comport with gardens as they do. Historical, anthropological, and biological significance is not of this type. Maybe the popularity in the eighteenth century of the informal 'English garden' did reflect, as historians often tell us, a commitment to sturdy British ideals of liberty, but that is not why figures like Horace Walpole so admired the genre. Nor, when I enjoy the prospect of an orchard from the refuge of a gazebo, has my reason for this anything to do with what had survival value for prehistoric hunters and gatherers.

The significance that the garden has for people need not, of course, be something that they are able easily, or even at all, to spell out. There is nothing untoward in the idea of structures of meaning which are implicit in people's lives and which it is the ambition of the philosopher—the phenomenologist, for example—to expose and articulate. To be sure, there is a problem here: how does one ensure that the significance of gardens as articulated by the philosopher corresponds to that which they have for the people who make and enjoy them? How, for example, might one establish the claim that the garden matters to people because it 'mediates' between various 'oppositions that define human experience', such as 'man and nature' or 'action and contemplation'? (Miller 1993: 57). But this is not a problem of an unfamiliar kind, and I shall be addressing it when, in my final chapters, I develop a proposal for answering our 'fundamental question'.

That question is a large one, I remarked, even when construed in the limited way just explained. Certainly it is a question that both ramifies into further questions and connects up with still broader ones. In part this is because gardens, some of them at least, are such complex and versatile places. In a much quoted passage, Eugenio Battisti reminds us, for example, that the Renaissance garden was 'a place of pleasure ... feasts, entertainment of friends ... sexual and intellectual freedom ... philosophical discussions, and a restorative for both the body and the soul.... [with] the function of a sculpture gallery, ... a horticultural encyclopedia *in vivo*, a centre of botanic and medical research. Finally, it is a perpetual source of moral instruction' (quoted in Strong 2000*a*: 14).

One thing this passage brings out—as, in a different idiom, do Andrew Marvell's garden poems—is that the significance of the garden cannot be restricted to the domain of the aesthetic. That the garden affords sensory pleasure and invites the exercise of taste is, to be sure, an important dimension of the significance that gardens have for many people, but not one that even begins to exhaust the

place that these same people afford to the garden within a wider
conception of 'the good life'. The 'repose...fair quiet...innocence...
[and] delicious solitude', for example, to which Marvell alludes in
'Thoughts in a Garden', are not aesthetic virtues of the garden—
not, at least, on currently favoured definitions of 'aesthetic': but they
are ingredients of the good life. No answer, therefore, to the question
'Why garden?' could be adequate that did not expose the significance of the garden in relation to an understanding of the good life.
This is a point that was appreciated by Gertrude Jekyll, for one, and
the reason why she wrote more on 'homely' than on 'great gardens':
for while the latter may be 'fine art' and the products of a 'master-
mind', it is the 'homely border of hardy flowers' that is more conducive to such aspects of the good life as 'happiness and repose of
mind' (Jekyll 1991: 91, 168). It is for a similar reason that this book has
no particular focus on 'great gardens', like Stowe or Versailles, and is
at least as much concerned with the 'homely', with the gardens that
more emphatically and intimately enter into our lives.

There is, to be sure, a rather breezy view—one that goes back at
least to Dr Johnson, for whom gardening was an 'innocent amusement', the 'sport' rather than the 'business of reason'—according to
which the question 'Why garden?' hardly calls for profound reflection. The story is told of a famous gardener visiting Charles Jencks's
Garden of Cosmic Speculation in Scotland and remarking that
he did not care about 'the meanings of gardens' provided they are nice
places to be in and look at. We garden and spend time in gardens,
the implication is, simply because these are enjoyable, amusing
things to do. But Tim Richardson is surely right to brand the famous
gardener's attitude as 'artificial', as a 'conceit' that appeals to an
'anti-intellectual strain in British culture, particularly when it comes
to gardens' (Richardson 2005: 8). A long tradition of garden literature
is replete with testimonies, of the kind I have already cited from
several authors, to the experience of gardens as pure and innocent,
rewarding and important, morally and philosophically instructive,
restorative and reposing. It would be philistine to ignore that
tradition, and itself a pretension to dismiss the testimonies of that

tradition as pretentious expressions of nothing more than the 'amusement' afforded by 'nice places'.

In the senses of the two terms explained by Ronald Hepburn when discussing the appreciation of nature, appreciation of gardens can be 'serious' as well as 'trivial'. A garden, like a mountain stream, may indeed be the object of 'pleasant, unfocused enjoyment'; but, like a mountain landscape, it may, for example, also be 'apprehended with a mysterious sense that...components deeply matter to us', even if we cannot say how (Hepburn 2001: 1, 10). And because gardens may be 'seriously'—which does not mean 'solemnly'—apprehended, the question of why we garden and why we live in and with gardens is itself a serious one: one, that is to say, for philosophical reflection. But that returns us to the puzzle of why, therefore, philosophy appears to have been so neglectful of the garden.

Philosophers in the Garden

One thing the literature is not short on is observations on the relationship between philosophy and the garden. One thinks, for example, of the theme, already alluded to by Battisti and going back at least to Epicurus, of the garden as a place peculiarly suited to the conduct of philosophical thought and discussion—a theme continued by John Evelyn in his 'endeavour to show how the aire and genious of Gardens...contribute to...philosophicall Enthusiasms' (quoted in Casey 1993: 154). One thinks, as well, of speculations on how the philosophical context of an age may be reflected in contemporary garden taste. Geoffrey and Susan Jellicoe's *The Landscape of Man*, for instance, includes in each of its historical chapters a sketch of the dominant philosophical mood—'delightful materialism' in seventeenth-century France, say, or 'philosophic liberalism and empiricism' in Germany and England during the same period (Jellicoe and Jellicoe 1995: 178, 192)—that formed part of the cultural backdrop against which the gardens of those countries were then designed.

But to reflect in these ways on connections between philosophy and gardens is not, *per se*, to reflect philosophically *on* gardens, and once we turn to the literature for reflections of the latter kind—for philosophy *of* gardens—we are indeed struck by the relative neglect of which several modern authors complain. Still, it is easy to exaggerate the complaint: neglect has by no means been complete. The degree of neglect is exaggerated, for a start, if attention is paid only to the writings of 'professional'—as it were, card-carrying—philosophers, those likely to figure in the canonical lists of philosophers found in potted histories of the discipline. Such lists will probably not include Virgil and Pliny the Younger, Albertus Magnus and Piero de Crescenzi, Evelyn, Pope, and Addison, or Karel Čapek and Hermann Hesse: but these are men who were philosophically trained and informed, and who brought their philosophical understanding to bear on the gardens they loved and, in several cases, created. The degree of neglect is also exaggerated if attention is confined to the Western philosophical tradition. For example, and as one might expect, the philosophical poetry of Japan is replete with philosophical reflection on the significance of gardens and of the things that grow or stand in them. Or consider the Buddhist tradition: both in the *suttas* of the Buddha himself and in such works as the *Theragāthā* (Songs of the Elders), there is discussion of the groves and parks in which early Buddhists passed much of their time as moral and intellectual resources—as places, reflection in and on which, cultivates understanding both of the virtues and of the nature of reality (see Cooper and James 2005: ch. 6).

It would be wrong, moreover, to imagine that authors who do figure on canonical lists of Western philosophers have had nothing to say about gardens. Shaftesbury, Kant, Hegel, Schopenhauer, Wittgenstein, and Heidegger may not have said very much, but they said something, often of the first interest. And it is not difficult, in some cases, to work out from their actual remarks in conjunction with their wider views what more these figures would have said about gardens if pressed. A striking instance, as we will see in a later chapter, is Heidegger's remarks on *bauen*—'building' or 'cultivation': these focus primarily on building, but it is not difficult to discern what Heidegger would have said about cultivation, that of gardens included, had he chosen this as his focus.

Finally, one should note that while the volume of professional philosophical writings on the garden over the last decade or so is, as I mentioned, modest, it is not negligible. Indeed, including as it does at least two substantial books (Miller 1993; Ross 1998), it can be described as modest only when compared to the explosively expanding volume of writings on other topics which has been characteristic of academic activity in today's climate of 'publish or be damned'.

Let's concede, though, despite the above caveats, that modern Western philosophy is guilty as charged of relative neglect of the garden. While I have urged that a philosophy of gardens should not be restricted to the aesthetics of gardens, it is nevertheless within the domain of aesthetics that one might above all have expected discussion of the garden to have been more prominent than it has been. There are, I suspect, at least four reasons why this has not been so. To begin with, gardens are typically and strikingly different in a number of respects from those artworks which have tended to be regarded as paradigmatic by philosophers of art and which have accordingly moulded, as Mara Miller puts it, the 'principal notions of aesthetics' (Miller 1998: 274). We will encounter some of these differences in the next chapter, but they certainly include the practical, utilitarian uses to which gardens, unlike paintings or sculptures, are typically put, and the high degree of dependence upon environmental factors—a lack of 'autonomy'—that gardens, again unlike paintings or sculptures, have. Second, and relatedly, gardens have suffered as a result of those complex contingent and historical developments, economic ones included, that by the nineteenth century had helped to favour what John Dewey called the 'museum conception of art'. This conception, by placing on a 'pedestal' products that could be conveniently exhibited in museums, galleries, or salons, at the same time marginalized the aesthetic significance of those activities—decorating and gardening, for example—more directly integrated into communal existence as 'enhancements of the processes of everyday life' (Dewey 1980: 5–7). Thus, while some gardens are often recognized as 'works of art', gardening rarely figures among 'The Arts', the primary focus, unsurprisingly, of the philosophy of art.

A third reason for aesthetics' neglect of gardens is surely the deprecatory attitude towards the garden of some of the founding fathers of the discipline, above all Hegel. For Hegel, gardening is an 'imperfect art', and gardens, while welcome if they provide 'cheerful surroundings', are 'worth nothing in themselves' (Hegel 1975: 627, 700). His main reason for this verdict is that the garden represents an uneasy mixture of—indeed, a 'discord' between—nature and art. Since, according to Hegel, aesthetics should be synonymous with the philosophy of art, the garden—in which art is badly muddied by nature—is not a proper or serious subject for the discipline at all. Not all aestheticians, of course, have followed Hegel in confining the remit of the discipline to art, and they speak happily of an aesthetics of nature. But if, for Hegel, the garden has too much of nature in it, then for many early—and indeed later—aestheticians of nature, such as Shaftesbury and Thoreau, the garden is too sullied by art to be an appropriate, or at any rate paradigmatic, object of natural appreciation. In recent years, indeed, among some radical exponents of environmental ethics, the garden—standing as it does at such a remove from 'wilderness'—is nothing to appreciate at all, aesthetically or in any other way. (In Chapter 3, however, we will encounter the familiar view that the 'fusion' of art and nature that gardens are alleged to be is a strong reason *for*, not against, paying them close aesthetic attention.)

A final factor contributing to aesthetics' continuing neglect of the garden reflects the degree to which the agenda of contemporary aesthetics has been set by 'late modern' developments in the arts—developments, however, to which the garden has largely been immune. A common perception is that, in contrast with painting, people no longer 'make statements' in the medium of gardening, so that in effect the art of the garden is 'dead', the corpse of a once live tradition whose true heir in contemporary culture is 'environmental art', the construction of 'earthworks', for example (Ross 1998: 192 ff.). While there may be some analogues in the gardening world to the abstracts, ready-mades, and 'conceptual' works that 'make statements' (usually about art itself) in the modern art world—statements that today's aestheticians dedicate a remarkable amount of energy to pondering—these have clearly not 'caught on' or transformed people's sense of what the garden is. Corpse-like or not, most gardening is happily undertaken without the ambitions of 'psychological and metaphysical speculation' (Adams 1991: 334) that have been so prominent in 'late modern' painting and that have so shaped the philosophical predilections of today's aestheticians. (There is a nice irony here: this reason for aesthetics' neglect of the garden, one suspects, is also a reason for the renaissance in popular enthusiasm for gardens noted earlier. What contemporary art galleries rarely offer people—beauty, delicacy, or grandeur, for example—gardens are still allowed to do.)

There is, I suggest, a more general reason for the neglect of the garden, not only in aesthetics, but in modern philosophy at large. As noted earlier, a prominent theme in writing on gardens since classical times has been the place of the garden in 'the good life'. For Pliny, a life largely spent in the gardens he himself laid out was 'a good life and a genuine one', and their cultivation belonged to his endeavour to 'cultivate myself' (Pliny 1963: 43, 112). For Joseph Addison, nearly two millennia on, the garden offers 'one of the most innocent Delights in humane Life', reflective of 'a virtuous Habit of Mind', above all for the opportunities it affords for that most characteristically human of capacities, the exercise of imagination—a pre-condition of wisdom, creativity, and insight (in Hunt and Willis 1988: 147). As these citations indicate, the good life with which the garden engages is neither the fun-packed life to which the expression popularly refers, nor the life of specifically *moral* rectitude. This does not mean that pleasures and moral probity are not ingredients of the good life: but what is intended by the expression is the fulfilled, flourishing, consummate human life that the Greeks called 'eudaimonic'—one in which, no doubt, there are many types of ingredient.

Reflection on the good life, on its ingredients, and on how to achieve it, was once the central concern of philosophy. Indeed, as Pierre Hadot reminds us, for many Greek thinkers, above all the Stoics, the title 'philosopher' referred in the first instance, not to a theorizer, an exponent of 'philosophical discourse' and doctrine, but to a person engaged in an 'act', an attempt through various 'exercises'—intellectual, spiritual, meditative, character forming, even physical—to live the good life (Hadot 1995: 266 ff.). Within modern philosophy, however, the issue of the good life, far from remaining central, has hardly been addressed. The reasons for this disappearance are various and complex, but they surely include the compartmentalization of professional philosophy into specialisms that deal only with restricted aspects of goodness (moral goodness in the case of ethics, for example) and the entrenchment of the attitude that, with the possible exception of moral judgement, views on the nature of the good life are purely a function of individual taste and 'choice', so that it is not a matter for 'reflection' to contribute to. But whatever the exact reasons, philosophical attention to the good life—and hence to the possible place of gardens in such a life—has atrophied in modern times.

That said, there has been a welcome revival over the last decade or two, in the form of so-called virtue ethics, of an older concern with the good life—a revival born from the sense that there are indeed aspects, other than narrowly moral ones, of the flourishing human life that are proper topics for philosophical reflection. (See, e.g., Annas 1993.) An equally welcome spin-off from this revival has been a handful of books examining ordinary human activities, such as eating and dressing, from the perspective of their significance within lives well led. We find, for instance, the author of a work subtitled *Eating and the Perfecting of our Nature* urging reflection, in a culture that has largely forgotten it, on 'what it *means* to eat', and the 'recover[y of]... the deeper meaning of eating' (Kass 1994: 231). Substitute 'garden' for 'eat', and the concerns of that author are visibly related to the 'fundamental question' that the present book addresses.

I began this chapter by giving a reason why the book is called *a*, not *the*, philosophy of gardens. Perhaps a further, rather deeper reason is emerging. Philosophy, as the ancients understood the term—as an 'act' or 'exercise'—does not permit references to *the* philosophy of… But it allows one to speak of a philosophy of gardens—or of eating, or of dress, or of much else: for this will be a way of trying to relate to and experience gardens (or food or dress) in manners germane to living well. In its own vicarious way— through the means of the pen or the PC rather than the spade or the trowel—a book may belong to that endeavour. It can be an attempt to attune people, its author included, to the significance of the garden and its relation to the good life. So, while much of the book belongs to what Hadot calls 'philosophical discourse', it is also an essay in what he, like the Stoics, would regard as philosophy in its 'existential' sense—as an 'exercise', a 'way of life', that 'philosophical discourse' attempts to articulate and render.

What is a Garden?

There are a number of bona fide philosophical questions that have loomed large in the modest revival of the aesthetics of gardens but with which, as I remarked earlier, I shall not be unduly concerned. Not only are they too close to similar and familiar ones asked about other artworks to raise novel issues, but answers to them—even if I had them—would not be germane, except here and there and in passing, to the theme of the book. The questions I have in mind are, at a pinch, subsumable under the umbrella one of 'What is a garden?'.

One way to hear that question is as a request for a definition of the term 'garden', in the classical mode of providing necessary and sufficient conditions for the term's correct application. But, while it is a reasonable ambition to provide such a definition, it is quite wrong to suppose that, without one being furnished, we do not understand the term and do not, in a perfectly acceptable sense, know what gardens are. We possess the knowledge that enables us to—and indeed, largely consists in the capacity to—distinguish gardens from those bits of the world that are not gardens. 'Garden' is an entirely familiar term, and nearly every English speaker knows what it means. Pressed to say what I mean by the term, my response would be 'The same as you who are pressing me mean by it—so you already know what I mean'.

It is also a mistake to think that philosophical discussion of the garden cannot proceed unless forearmed with a (classical) definition of 'garden'. A main purpose of such a definition is to provide guidance on how to handle 'grey' or 'marginal' cases—ones where everyday, implicit understanding of a term does not yield agreed and clear-cut intuitions as to whether the term applies or not. Now there may indeed be 'marginal gardens'—ones in which, say, nothing grows, or the matchbox-sized 'Zen Garden' I recently bought, or a *jardin trouvé* (a bit of land dubbed a garden by someone from the 'Art is what *I* call art' school of thought). But such places, even if we decide—with or without the help of a definition—to count them as gardens, are not, of course, of the kind that over the millennia have inspired the appreciation and engagement of human beings. And it is with the kinds that have done this that the present book is concerned. (Here's an analogy: for the purpose of certain discussions about music it may be helpful to provide a definition that dictates whether John Cage's silent work '4½', or the chimes of Big Ben, constitute music—but not, I suggest, when the focus of the discussion is why music matters to people.)

For these reasons, I am content, if a definition must be given, to go along with ones that strike me as capturing reasonably well people's everyday implicit knowledge of what gardens are. Not all definitions that have been proposed do that, of course: not, for example, the *Shorter Oxford English Dictionary* one—'an enclosed piece of ground devoted to the cultivation of flowers, fruit or vegetables'. An unenclosed place, full only of trees, shrubs, and grass, might still unquestionably be a garden. Much closer is Mara Miller's stab: 'any purposeful arrangement of natural objects ... with exposure to the sky or open air, in which the form is not fully accounted for by purely practical considerations such as convenience' (Miller 1993: 15). Critics have been quick to point out that this definition will not cover the unusual creations of some soi-disant garden designers— totally enclosed ones, for example, or ones with synthetic substitutes for natural objects. And maybe Miller is rash to describe all such creations as being garden 'imitations' or gardens only in a 'metaphorical sense'. It is true as well that the definition, as it stands, does not allow for distinguishing gardens from, say, public parks and motorway recreation areas that one might hesitate to call gardens. But as a 'working definition' that reasonably approximates an everyday understanding of what gardens are, it is one I am content to proceed on the basis of, amending it if and when this becomes germane to my task.

It is tempting, but not quite accurate, to describe a definition like this as articulating the 'stereotypical' concept of the garden—as covering just what nearly everybody would immediately, or with only the slightest prompting, recognize as a garden. In fact, though, it covers—and rightly so—gardens that are not 'stereotypical'. A lorry-driver, delivering a load to the house at Lunuganga of the great Sri Lankan architect and garden designer Geoffrey Bawa, took an admiring look at the very 'natural', 'informal', and almost flowerless garden, and then remarked, 'This would be a lovely place to have a garden' (Robson 2002: 240). Cart- and carriage-drivers in eighteenth-century England, I imagine, might have said the same about the equally 'natural', flower-free landscape garden of 'Capability' Brown at Blenheim. Lunuganga and Blenheim are not 'stereotypical' gardens—not, as it were, the sort a child, or indeed an adult, might draw when asked to draw a garden. Yet they are, of course, gardens—'atypical', if you wish, but gardens none the less. A similar point could be made about a very different type of 'atypical' garden—the 'dry landscape' (*karesansui*) of Japanese Zen tradition. When, with pardonable paradox, it is said that the famous treeless, flowerless, birdless garden at Ryōanji in Kyoto is 'disconcerting through its not really being a garden' (Berthier 2000: p. vii), the point, I take it, is to recognize this place as an 'atypical' garden.

Not every act of defining a term takes the classical form of
providing necessary and sufficient conditions for its application.
When R. S. Thomas opens his poem 'The Garden' with the lines
'It is a gesture against the wild, / The ungovernable sea of grass', or
Miller characterizes the garden as 'an attempt at the reconciliation
of opposites which constrain our existence' (Miller 1993: 25), they are
exposing what they see as some deep aspect of the garden of a kind
that would hardly figure in a dictionary entry for 'garden'. And
when the American garden designer Thomas Church speaks of
'the term "garden" having changed its meaning'—from that of a
'place to be looked at' to that of one 'designed primarily for living'—
he is not reporting on a change in the semantics of the term, but
drawing attention to an important shift in the way that people
experience and relate to gardens (Church 1995: 32). With the definition of a garden in these broader, non-classical modes of definition, I
shall indeed be concerned in this book. But such definitions, far
from having to be set out before discussion can continue, are of a
kind that the whole discussion should be seen as groping towards.

The question 'What is a garden?' need not be heard as a request
for a definition. Instead, it may be a request to address various, related
issues of an ontological kind: What kind of being does a garden
have? What sort of entity or object is a garden? Is it simply a complex
physical object, say? When, and as a result of what sorts or degrees
of change, does a garden cease to be the garden it was?, and so on.
Questions of this type resemble, of course, ones asked in the ontology
of art about art objects: Is a painting simply a physical object? What
sort of existence does a symphony, as distinct from performances
of it, have? Does a block of marble cease to be the same work once
removed from the Greek temple and relocated in a millionaire's
bedroom?, and so on. Now it may be that some such questions, when
asked about gardens, raise fresh problems or call for answers uniquely
appropriate to gardens. But I doubt it. It has been suggested, for
instance, that gardens are unique among artworks in having to
occupy particular locations—so that the same garden could not,
over time, change its location. The identity criteria for gardens,

therefore, are quite different from those for, say, paintings or vases. But we talk these days, and without any obvious tension, of a garden at the Chelsea Flower Show being moved, after the show, to somewhere in Yorkshire or even Kuwait. And anyway it is hardly self-evident, even if it were true that a garden cannot change its location, that this would be a feature unique to gardens, as the above query about the marble statue moved from its temple indicates.

Even if I'm wrong in supposing that the ontological problems of gardens are not of a special sort, they are not ones, for the most part, that are relevant to the kind of enquiry I am engaged in. While, for example, the fact that a garden changes in salient respects with the seasons or the moods of the day will indeed be germane to explaining the significance that gardens have for people, the fact—if it is one—that a garden which changes in respects X, Y, and Z is no longer one and the same garden will not.

There are, though, two issues of an ontological type about which I want to say a little at this stage, for they will bear on my enquiry at certain points. The first concerns the *counting* of gardens. The concept of a garden is a peculiar one in that, while I rightly speak of owning only one garden, I also speak naturally, as I walk round our house, of moving from the Japanese garden to the cottage garden, and from there to the 'Mediterranean' garden. So haven't I got three gardens, not one? And what about the many people with a front and a back garden? Do they have one garden, bisected by a house, or two gardens? Again, there are 'great' gardens, like Sissinghurst, Hidcote, or Highgrove, that are described as containing different 'rooms'—discrete gardens that constitute what is none the less referred to as *a* garden. This phenomenon of 'gardens-within-a-garden' is an interesting one, not because it really matters much how we decide to speak of Sissinghurst, say, as six different gardens—rather than as *a* garden divided into six 'garden rooms'—but because the appeal and significance of many gardens, 'homely' as well as 'great', may owe something to it. At the very least, one would like to explore why people—even when they strive for the 'unity' which garden designers like Sylvia Crowe (1994: 94) and Russell Page (1995: 13) deem

Plate 1. Garden 'rooms', Kailzie, Scotland

imperative—also find it important, as indeed do Crowe and Page, to create 'variety' in the form of relatively discrete and 'autonomous' areas within the whole of their gardens.

Second, I want to say something about the strangely well-subscribed claim that gardens are not, primarily at least, physical entities, but 'virtual' or 'ideal' ones—less because I find the claim plausible than because some, at least, of the thinking that inspires it is important and relevant to my later discussions. Stephanie Ross writes that we should 'distinguish (1) the physical garden, a particular chunk of [say] Surrey, from (2) the world of the garden, a virtual world' (Ross 1998: 179). Despite the terminology, this is not the claim, often made in connection with paintings, that we must distinguish the physical artwork from 'the world of the work', in the sense of the world, say ancient Greece, which the work represents. Certain gardens, of course—Chinese ones, for example, that re-create mountain ranges—are designed to represent a 'world', but Ross's distinction is intended to be entirely general, applying as well to non-representational gardens.

Some of the arguments that lurk behind the distinction are unpersuasive. One suspects, for example, that there lurks the bizarre thought that only the physical—only what features in physicists' accounts of the world—is truly 'real', so that if gardens are not 'merely' physical, they can only be 'virtual', 'ideal', or even 'illusory' (Miller 1993: 122). And certainly the following *doppelgänger* argument, familiar once again from discussions of other artworks, gets invoked: a garden cannot be identified with a physical entity, since it may be completely indistinguishable, in physical respects, from a piece of land that is not a garden, but a natural place that chances to look as if it were. But all that this argument establishes is that the criteria for identifying a place as a garden, notably that of its having a human input, are not the same as those for identifying a place as natural. This hardly warrants talk of 'two worlds', a physical and a virtual one. (I am not tempted to describe the stone paperweight on my desk as 'virtual' simply because it bears the mark of human intervention, polishing.)

But perhaps the main reason behind the distinction emerges when, for instance, Tom Leddy—during a discussion of Ross's claim—writes that 'if you knew [the place] was a human phenomenon you would look at it differently. You'd wonder what the meaning was', for example, in a way you wouldn't in the case of a 'purely natural phenomenon' (Leddy 2000: 2). The thought, I take it, is this: to describe, experience, and engage with a place as a garden is to do so in ways quite different from those appropriate to a 'merely' natural place, and certainly from the ways in which a physicist or botanist or mineralogist might describe or experience places. That thought, which I shall be taking up at various points—for example, in Chapter 2, when comparing and contrasting garden and nature appreciation—is true and important. Consider, in this connection, Karel Čapek's charming description of the ways in which a gardener may experience the soil he deals with. Having identified, rather as a mineralogist might, the acidity and stoniness of the soil, the gardener proceeds, as the mineralogist doesn't, to distinguish among soils 'as airy as gingerbread, as warm, light and good as bread . . . as fat as bacon, as loose as cake . . . [as] greasy, [and] cloddish'—distinctions inseparable from perceptions of the soils as 'beautiful', 'noble', 'ugly', 'sterile', and 'malicious', or whatever. These are not at all distinctions that would be made by the physicist or the mineralogist who confines himself to describing and observing the soils as so much 'unredeemed matter' (Čapek 2003: 126 ff.).

As these remarks might suggest, however, the truth to the thought indicated by Leddy's comment does not at all require a counterintuitive and obfuscating 'two worlds' distinction. There are not two soils, the one examined by the mineralogist and the one responded to by Čapek's gardener: there's simply the soil—but described, experienced, and engaged with in quite different ways. Gardens are not 'virtual' or 'ideal' places, somehow or other connected with physical 'chunks' of land: they are simply places, albeit ones that invite a rich and varied range of description, experience, and comportment.

I have spent some time in this chapter explaining what I shall not be discussing in any detail in this book. Readers might therefore welcome a brief preview of what I shall be discussing.

In Chapters 2–3, the part of the book most orientated towards aesthetic matters, I explore the relations between garden, art, and nature appreciation. The first of those chapters examines and rejects the views that gardens matter to us, in the final analysis, for the same reasons that artworks and natural landscapes respectively do.

The second examines and also rejects the more subtle view that gardens are appreciated as 'fusions' of art and nature.

Even if the views considered in those chapters were more compelling than they are, they would at most illuminate some aspects of the significance and importance that gardens have for people. In Chapter 4, I consider some of these further aspects, focusing on ways in which gardens engage with people's lives and contribute to their living well. In this and the following chapter, 'the good life' is not understood with any particular reference to morality: but in Chapter 5, I try to articulate respects in which gardening and other modes of engagement with the garden contribute to recognizably ethical dimensions of the good life. At the same time, I consider and respond to a number of moral concerns that some people have about the enterprise of gardening.

In Chapters 2–5, as indeed in the present one, there will have been a good deal of talk about the meaning or significance of gardens. But it is not until Chapter 6 that I try to lend some order to such talk and to distinguish among the many modes of meaning that a garden might possess. Not all of these have much bearing on the 'fundamental question' of why gardens matter, but one at least does— that of the garden as signifying something about the relationship between human beings and the world through 'exemplifying' or 'embodying' that relationship in a salient way.

In Chapter 7, this notion of significance is taken up in detail, and it is argued that the garden is an 'epiphany' of the relationship between human creative activity, the world, and the 'ground' of the world. In the final chapter, after recapitulating earlier discussions in the light of the conclusions of Chapter 7, I suggest that it is in virtue, ultimately, of being an epiphany that the garden belongs to the good life and is of fundamental significance to people.

2 ART OR NATURE?

Two Models

When seeking to explain some relatively unexplored phenomenon—in our case, the 'serious' appreciation of gardens—it is tempting to assimilate, or even reduce, what needs explaining to something that, we feel, is already familiar and explicable. We seek models, one might say, for what we want explained. Gardens—'typical' ones, at least—are natural places that have been transformed by creative human activity, and contain natural items, such as flowers, chosen, placed, and organized at least in part in accordance with aesthetic considerations. This truism suggests two obvious models for garden appreciation: those of art and nature appreciation respectively. In this chapter, I explore and reject attempts to assimilate garden appreciation to either of those two types. In the following chapter, I consider the more subtle view that garden appreciation somehow fuses, or factors out into, art *and* nature appreciation. But it will help us to understand and assess that view if we first consider the either/or options: either art or nature appreciation is the model for garden appreciation.

Several preliminary remarks are in order. First, I am using the term 'appreciation' in a broad way, to cover such 'pro-attitudes' as enjoying, taking pleasure in, loving, admiring, and finding significance or meaning in. As I use it, then, appreciation is not exclusively to do with the exercise of taste. Second, I use the term 'nature' in what might be called 'the David Attenborough' sense: the natural environment—those stretches of the world, including meadows and hills, that are relatively unaffected by human intervention, along with ingredients of those places, including flowers and rocks, whose existence is similarly independent of human artifice.

Since, third, I shall reject the assimilation of garden appreciation to that of art or nature—or a fusion of the two—I am fortunately spared the task of giving an account of why art and nature are appreciated by us, why they matter to us, even if occasional remarks on this will be pertinent. I want to say that whatever remotely plausible accounts are given of why we appreciate art and nature, these do not provide models for an account of garden appreciation. Finally, I shall ignore the claim, an implausible one anyway, that there are no differences, *au fond*, between art and nature appreciation—that, in particular, nature itself should be viewed 'as if it were art': a view, one suspects, with its roots in the venerable idea that nature really *is* art, 'the art of God', as Sir Thomas Browne called it. (See Budd 2002: 119 ff. for criticism of the claim.) So there are indeed two models, not just one, of garden appreciation.

It is essential to distinguish my question, concerning the assimilation of garden appreciation to other types, from a number of related questions of a normative or evaluative kind, ones that especially preoccupied the eighteenth century, not least because of their perceived bearing on the obsessive debate—still with us today—over the respective merits of 'formal' or 'regular' gardens and 'informal' or 'natural' ones. (This debate, as we shall see in Chapter 5, was a moral as much as an aesthetic one: with Arthur Schopenhauer, for instance, castigating French gardens that 'forcibly imposed' forms on nature as 'tokens of her slavery' (Schopenhauer 1969: 405).) To begin with, mine is not the question of whether art is 'superior' to nature, or vice versa—whether, say, Chatsworth is a 'paradise' in comparison with 'Nature's shames and ills', as Charles Cotton maintained in 1681, or whether, conversely, the artificial 'magnificence' of Versailles is nothing in comparison with 'the sportive play of nature in the vale of Keswick', as Arthur Young insisted in 1770 (quoted in Andrews 1999: 68, 71).

Nor, more importantly, is mine the question of whether gardens should look like and resemble nature, and if so what sort of nature—very 'rude wilderness' or only a 'very politely rude' wilderness, as Simon Schama puts it when reflecting on the great eighteenth-century debates (1995: 539). It is perfectly possible to hold that, while the garden should indeed look like a natural place— even to the point of being mistaken for one by the ignorant—it is nevertheless as a work of art that the garden must be appreciated. The point might be, for example, that gardens—like those discussed in the Japanese classic and oldest of all treatises on the garden, the *Sakuteiki*—while 'distillations' and partial 're-creations' of the natural world, are to be enjoyed 'as works of art', not least for their skill in

Plate 2. Gertrude Jekyll, garden on Holy Island, Northumberland

artfully distilling and re-creating nature (Takei and Keane 2001: 42). Or the point might be, as one doughty champion of the 'natural look', Gertrude Jekyll, indicates when announcing her intention to 'make a beautiful garden-picture', that appreciation of a garden demands just the sort of attention to 'form and proportion' and to colour combination that appreciation of paintings does (Jekyll 1991: 48, 72). Conversely, it is possible to hold that garden appreciation is essentially nature appreciation, even though gardens should *not* look like anything we actually encounter in the natural world. The best gardens, it has been held, are 'nature perfected', 'improved', or 'methodised', as Pope put it (1994: 6), and hence are to be enjoyed or admired as approximations, not to the empirical world, but to an 'ideal' nature from which 'rude' nature marks a 'fall'. A variation on that characteristically Renaissance conception of the garden is, as Bouvard and Pécuchet seem to have thought, that we are affected by gardens, not for their resemblance to 'the rolling greenery' of visible landscapes, but because they call to mind 'what could not be seen ... what was beneath the earth' (Flaubert 1976: 92).

The Garden as Art

In this section, I consider the issue of whether garden appreciation may be assimilated to that of artworks, above all of paintings, the most favoured analogue to gardens. This issue should be distinguished from ones with which it might be confused. It is not, for example, the issue of whether, as one advocate of the 'picturesque', Richard Payne Knight, urged, 'the whole pleasure' in viewing gardens derives from 'associating' them with landscape paintings (in Hunt and Willis 1988: 350). Doubtless, over the centuries, there has been much associating and mutual borrowing between gardening and painting: after all, many famous garden designers have been painters—Jekyll, for example, and of course Claude Monet. But associations with, and borrowings from, paintings do not establish that the model for appreciation of a garden is appreciation of a painting.

More importantly, our issue is not whether gardens—some of them, at least—may be legitimately called 'works of art'. It seems obvious to me that some of them may, and that some garden designing may therefore be called 'art'—in a more serious sense, moreover, than that in which we can speak of golf or love-making as an art. Golfing and love-making, though skills that allow for grace, delicacy, and originality, are not undertaken with the intentions, nor do they have the end-products, that qualify them for being arts in the sense that painting and musical composition are. Putts and caresses, unlike making a parterre or designing a pond, do not have the right kind of 'provenance', to use David Davies's (2004) helpful term, to count as constituent activities of an art-form.

Even if garden design does not usually feature on lists of 'The Arts'—ones which have anyway been drawn up as the result of contingent, historical, and social factors—it is counter-intuitive to deny certain gardens the status of art. Moreover, the garden does reasonably well on some of today's favoured definitions or characterizations of art. For example, 'functionalist' ones which emphasize such purposes of art as the presentation of objects for aesthetic attention should be able to accommodate some gardens without difficulty. Gardens can be accommodated too, I suspect, by 'institutional' definitions, which identify artworks in terms of what is deemed to be an artwork by or within an appropriate 'institution', such as the so-called Artworld. It is not simply that Artworld and Gardenworld intersect—almost as much today as in eighteenth-century England or eleventh-century Kyoto—as a result, for instance, of many figures participating in both worlds. In addition, Gardenworld is replete with figures of the same kind that populate Artworld—creative designers, craftsmen, critics, connoisseurs, and so on—and with similar organizational features—competitions, shows, dissemination through photographs, and so on.

The legitimacy of calling some gardens works of art, and some gardening an art, does little or nothing, however, to secure the claim that garden appreciation is a type of art appreciation. For that claim, recall, is intended to be *explanatory*, and it can only be this if garden appreciation is being assimilated to that of artworks *other than* gardens, such as paintings. The claim gathers whatever explanatory force it has from the attempt to show that garden appreciation is closely akin to modes of appreciation which we think we *already* understand. Clearly, one does nothing to further that attempt merely by remarking that the elastic term 'art' may legitimately be applied to gardens.

On this matter, champions of the assimilation of garden appreciation to that of art have been clear: for their point has been that the former *can* be assimilated to appreciation of, say, painting or sculpture. I have already mentioned Jekyll's reference to the gardener as making 'garden-pictures': elsewhere she writes that 'planting ground is painting a landscape with living things' and ranks, for that reason, among 'the fine arts' (Jekyll 1991: 160). Pope bluntly stated that 'all gardening is landscape painting,' and it has been said of the eighteenth-century designer William Kent that he was the first to have 'conceived the approximation of gardens to painted landscapes' (both quoted in Ross 1998: 90). Actually, Kent was hardly the first: the seventh-century CE Chinese flower garden, one reads, was often 'considered as landscape painting in three dimensions' (Wang 1998: 17), while 'the art of Japanese gardeners was always' considered 'akin to painting, particularly to ink drawings' (Hrdlička and Hrdlička 1989: 12 f.). (Readers might expect me to add to this list some citations from advocates of 'picturesque' gardening: but while that term was indeed used by John Gilpin, around 1770, in order to emphasize the garden's affinity to paintings, it soon came to denote, in the writings of Knight and Uvedale Price—admirers of the paintings of Salvator Rosa—a taste for gardens that emulated 'rough', 'rude', and 'sublime' nature of the sort depicted by Rosa.)

Painting, as remarked, is not the only art to which gardening has been assimilated. Sculpture, and even poetry, theatre, and ballet have had their champions as well. But if the painting model of garden appreciation is not persuasive, as I now want to argue, these other candidate models are unlikely to be found more so.

Anything, we know, is similar to and may be compared with anything else in any number of respects. Give me half an hour, and I could come up with a long list of ways in which enjoyment of a garden resembles that of a football match or a tasty dinner. Comparisons of gardens with paintings need to be serious and interesting to secure the claim that appreciation of the former may be assimilated to that of the latter. In my judgement, the comparisons frequently drawn between garden and art appreciation are too exaggerated, too lacking in relevance, or too superficial—sometimes a combination of these—to secure that claim.

By exaggeration, I have in mind such statements as that, when someone walks along the path of a Chinese garden, 'scenes will unfold ... *much as if* a handscroll were being unrolled' (Wang 1998: 39; my italics), or—to switch to another art—that Stourhead, with its 'emblems' of Aeneas's wanderings, is *'almost literally* a poem' (quoted in Ross 1998: 66; my italics). I do not find that what it is like to enjoy a garden I am walking through is much like what it is like to look at a handscroll, let alone to read a poem. (This is not to deny, of course, that there has been an important and intended interplay between garden design and the arts of painting and poetry.) By lack of relevance, I have in mind, for instance, the point that some gardens (such as Stourhead, once more) aim, like many paintings do, to represent places, people, events, or whatever. But most gardens do not do this, and in those that do, many or most of their ingredients— particular bushes, say, or patches of lawn—do not. There may be *cognoscenti* for whom it is the representational properties of gardens that appeal exclusively: but they are hardly typical of garden lovers, most of whom neither seek nor especially care about whatever representational ambitions a garden may have. Monet liked some of his paintings of his lily pond at Giverny because they were 'the mirror of water whose appearance alters at every moment' (quoted in Holmes 2001: 42), but he didn't admire the water itself because it was the representation, the mirror—in the same sense as the paintings were—of anything beyond itself. By superficiality, finally, I am thinking, for example, of the idea encapsulated in Jekyll's metaphor of garden-pictures. This metaphor, as she intends it, primarily encourages us to assimilate admiration of the colour combinations we sometimes find in gardens to that of the colour schemes which some painters compose. For a start, this exaggerates the phenomenological similarities between viewing 'drifts' of flowers and areas of a canvas: after all, it would be both difficult and peculiar to ignore the fact that, in one case, the colours are those of (actual) *flowers*. But the idea also ignores the contexts or environments in which we view gardens and paintings respectively. In the one case, we view the colours against a background of, say, trees and distant hills; in the other, as parts of something hanging on a white gallery wall. Such is the impact of context on our perception that there is little reason to think that the colour combinations which impress us in certain gardens coincide with those we admire in various paintings. Once properly examined, Jekyll's comparison looks to be both superficial and exaggerated.

It is not hard to identify the general reason why assimilations of garden to painting appreciation tend to be exaggerated, irrelevant, or superficial. Gardens are *places* whose ingredients, typically and significantly, are living, growing, or other *natural* things. Paintings are not places, nor do they (physically) contain living, natural items. It is worth setting out in some detail the salient differences of appreciation that are consequent on this obvious general difference between gardens and other artworks. (Several of the points that follow are well made in Mara Miller's *The Garden as Art* (1993: ch. 4), but in the context of explaining why aestheticians have not regarded gardens as pukka works of art, rather than the present one of resisting the assimilation of garden appreciation to that of (other) artworks.)

To begin with, there is the familiar point, attested to by many lyrical descriptions, that in experience of a garden all of the senses may be engaged—sight, hearing, touch, smell, even taste—in a way they are not in the presence of paintings. (Some avant-garde artworks, of course, aim to engage several senses, but appreciation of these—which relatively few people ever encounter anyway—could hardly serve as the model for something as venerable and widespread as garden appreciation.) Nor, one might add, are the senses separately engaged, in the sense that one could abstract a visual pleasure, say, or a tactile one from the *gesamt* experience. I wouldn't enjoy the feel of a wet stone beneath my bare feet in a Japanese garden in just the way I do unless I could see its glistening dampness studded with the matt green of the moss.

A second, equally familiar point is that, as the gardener Christopher Lloyd expressed it on a television programme, 'the garden is the most impermanent art...changing all the time'. Actually, there are three points here. The first is that the garden is peculiarly sub-ject to physical change: flowers grow and die, trees shed their leaves, and so on, and even within a single day rain, snow, or a blistering sun can bring about salient changes. 'Sissinghurst', it's been said, 'cannot be visited twice: it has always in the meantime, become something different' (Moore *et al.* 1993: III). Second, gardeners can themselves make very considerable changes to a garden—altering its layout, introducing new species, and so on—yet the garden remain the same one. Unlike a painting, a garden, as it is sometimes put, is never 'completed'. Third, the phenomenal aspects of a garden are continually and often strikingly changing, due to alterations in the conditions of perception, notably that of light, even when no (major) physical alterations, natural or man-made, are taking place. Sissinghurst, one might say, can't even be *viewed* twice: it has always, in the meantime, changed its appearance. The products of painting and the plastic arts are not 'impermanent' in any of the above ways, although they may of course be impermanent in other ways: they can deteriorate, for instance. (An exception, once more, needs to be made for avant-garde works which are either designed to undergo change or come to incorporate unplanned alterations, as in the case of Marcel Duchamp's *Large Glass*, which he exhibited replete with the dust the work had gathered in a warehouse.) Nor, one might add—a consideration which will assume more significance in Chapter 7—is the impermanence of gardens marginal to their appreciation, let alone something to rue and to envy the other arts for not having to cope with.

Two important further points are related by their attention to the physical—geometrical, one might say—relations of the observer

to gardens and paintings respectively. The historian of gardens
Marie-Louise Gothein writes of the landscape designer Humphrey
Repton that he 'free[d] himself from the exaggerated idea of a simi-
larity between painting and landscape gardening' once 'he had laid
his finger on the difference between them, caused by the constant
alteration in the [garden] spectator's point of view' (quoted in Howard
1991: 127). Typically, we do not just stare at a whole garden from a
window or a terrace, but look at it as we move around or through it,
or when otherwise actively engaged—taking a drink out on to the
lawn, say, or while watering a bush. Unlike the case with paintings
and even sculptures, there are no privileged points from which to
view and otherwise experience a garden. Some formal gardens, to
be sure, with their *alleés* and focal objects, invite certain views: but
one would hardly criticize or find eccentric the stroller who looks at
the garden from perspectives not invited by the designer—not in
the way in which one would be perplexed at a gallery visitor who
insists on looking at the paintings from the floor or the ceiling.

This already suggests the further point that gardens are not viewed,
as Arnold Berleant puts it (1993), as 'framed' objects standing there
before us in the manner of a painting on a wall. But at least two more
considerations are intended by Berleant's point. To begin with,
the person appreciating a garden, even when immobile, is typically
in, surrounded by, what he is appreciating: at no given point in time
is the garden *an* 'object'—a discrete thing taken in at a glance—for
his delectation. Furthermore, even when he is viewing the garden as
a whole, from a terrace, say, he is not doing so in the manner of a
person focused solely on a canvas clearly demarcated by its frame.
Horace Walpole famously remarked of William Kent that, by replacing walls and fences with ha-has, he 'leaped the fence, and saw that
all Nature was a garden' (quoted in Hunt and Willis 1988: 313). But even
when the walls or fences remain in place, the point is not affected.
It is not simply that the walls themselves—ivy-clad or espaliered,
perhaps—may belong to the garden rather than simply enclose it.
Additionally, when a person views the garden, he quite properly
takes in the sky and the 'borrowed' scenery of the surrounding

land- or townscape, and attends, as it gets put, to the 'genius of the place' where the garden is located. The walls do not therefore define or dictate what is and should be experienced—not in the way the frame of a painting does. (Incidentally, it is anyway unclear what counts as viewing the garden as a whole: as a three-dimensional place, there will always be parts of the garden invisible—occluded by trees, for example—from any given viewpoint.)

The final reason for resisting the assimilation of garden to art appreciation stems from the many practical and 'utilitarian' uses to which gardens have always been put. The medieval garden, for example, while likely to contain a small ornamental herber, was mainly given over to the 'kitchen or utilitarian garden' (Landsberg n.d.: 28); while the modern garden, to recall Thomas Church's words, is increasingly 'designed primarily for living as an adjunct to the functions of the house'—for eating, swimming, playing in, and much else. Some philosophers conclude that these practical, utilitarian dimensions of the garden are enough to disqualify it as an object of aesthetic appreciation in what they see as the 'traditional' sense of 'disinterested' or 'contemplative' appreciation. (They usually go on to add, 'So much the worse for the tradition!'. See Miller 1993: 98; Berleant 1993: 230 ff.)

But that is a contentious conclusion, and one which may betray a misunderstanding of the terms 'disinterested' and 'contemplative' as intended by the main representative of that tradition, Immanuel Kant. A less contentious conclusion is that, because of its practical aspects, appreciation of the garden is not to be assimilated to that of other, paradigmatic artworks—for these do not have, *qua* artworks, such utilitarian uses. Someone who defends the assimilation might argue that we can abstract away from our appreciation of a garden whatever pertains to its practical functions, and then ignore these in favour of the garden's artistic aspects—rather as we might ignore a painting's serving to hide a damp patch on the wall and focus on it solely *qua* a work of art. In response to that suggestion, I can do no better than quote two passages that nicely demonstrate the artificiality, indeed impossibility, of any such abstraction. George Eliot, recalling an 'old-fashioned' garden from her childhood, writes that there was

...no finical separation between flower and kitchen-garden there; no monotony of enjoyment for one sense to the exclusion of another; but a charming paradisiacal mingling of all that was pleasant to the eye and good for food... you gathered a moss-rose one moment and a bunch of currants the next; you were in a delicious fluctuation between the scent of jasmine and the juice of gooseberries. (Quoted in Wheeler 1998: 321)

Eighty years later, the gardener and author William Bowyer Honey wrote:

It is a familiar experience to find one's greatest aesthetic enjoyment... in something incidental, the by-product of another activity... In many gardens... planted for a practical utilitarian purpose, such experiences are very precious, and the joy taken in beauty of form and colour may be all the keener for its incidental character. (Quoted in Wheeler 1998: 232)

To the observations of these authors, one might add one made by Thomas Church: while gardens may be 'adjunct to the functions of the house', the designer should aim to create a sense of 'repose', 'peace and ease', and of being 'at home in [one's] surroundings' (Church 1995: 33, 49)—a sense, moreover, that cannot be isolated from recognition of and confidence in the garden's smooth efficiency in fulfilling its practical functions. A person will not experience that feeling—an 'aesthetic' one, if you wish—of ease and at-homeness which belongs to a rounded appreciation of his or her garden who is constantly worried about how to maintain areas for the children to play in or friends to sip cocktails in.

The objections to the assimilation or reduction of garden appreciation to art appreciation elaborated over the last few pages are all consequent, in part at least, on the truism that gardens are places that both contain and are affected by natural, including living, things. This is why gardens engage all the senses, are 'impermanent' in various respects, eschew privileged viewpoints, are not 'framed', and lend themselves to the practical purposes they do. In all of these ways, gardens typically differ from other works of art, notably paintings, in the types of appreciation they invite. The presence of natural things in or around (or above) gardens has unsurprisingly prompted some people to argue for the assimilation of garden appreciation to that of nature. That is the topic for the next section. I close this one by briefly responding to a worry that some readers may have been experiencing.

Haven't I, these readers may worry, been looking at the wrong arts? Aren't there arts, so far barely mentioned, appreciation of which may be a model for garden appreciation? The two candidates I imagine them proposing are environmental art and architecture. Neither proposal, though for different reasons, is persuasive. To see why, we need to remind ourselves of what the assimilationist is trying to do—which is not simply, or mainly, to draw attention to similarities between the garden and some other kind of artwork, but to *explain* our appreciation of gardens in terms of its affinity to that of other artworks.

In the light of that reminder, let us first consider environmental art—the 'earthworks' of Michael Heizer in Nevada, for example, or the landscape installations of Nancy Holt in Utah. The questions of the 'likeness' of such works to gardens, and of whether—as Ross (1998) and Leddy (2000) maintain—these represent the true continuation of eighteenth-century garden art, need not be addressed. For whatever the answers, it is apparent that environmental art is too new, and known to too few people, for appreciation of it to be the already familiar kind to which, in the attempt to explain garden appreciation, the latter could be assimilated.

Architecture does not, of course, encounter the problem of unfamiliarity, and it may well be that there are striking similarities between the reasons why certain gardens matter to people and why certain buildings do. The problem with invoking architectural appreciation in the present context is that, instead of illuminating our 'fundamental question' about gardens, it invites a very similar question itself. Granted that some buildings are 'works of art', as some gardens are, the question remains why many sorts of building—including ones with no pretension to be artworks—have such significance for people. It is no more obvious than it was in the case of gardens that the answer to this question can be in terms of why artworks matter to us—rather than in terms, say, of 'a sense of place', or, as in Gaston Bachelard's *The Poetics of Space* (1994), of 'intimacy' and the importance of 'nests'. Indeed, it may sometimes be helpful when trying to explain the significance of a building to invoke our appreciation of gardens, rather than vice versa. Maybe the rooms in certain buildings have the significance they do for people because they are redolent of certain gardens. If the metaphor of an 'outdoor room' may usefully be applied to some gardens, then why not apply the metaphor of an 'indoor garden' to some rooms?

The Garden as Nature

The objections to the assimilation of garden to art appreciation were all due to the fact that gardens are places which contain and are affected by, and open to, natural things or processes. Some of those who, rightly, attend to this fact have then proceeded to assimilate garden appreciation not to that of (other) artworks, but to that of nature itself. 'Nature' and its cognates are, of course, elastic and ambiguous terms, and not a few debates that have raged among gardeners betray equivocation over these terms. When, for example, William Robinson, the nineteenth-century champion of 'the wild garden', argued that it was natural to stock one's garden with plants introduced from abroad, his points were that one was thereby 'naturalizing' these foreign natives and entering into a less parochial 'communion with nature' (Robinson 1979: 9). In objecting to such introductions, however, his many critics have usually meant that it is unnatural to grow plants that are not ecological natives of one's country or parish. Again, some debates reflect the different uses of 'nature' to refer now to the natural environment that is visible to us, and now to 'the essential reality underlying all things' which, according to Monet's friend, Georges Clemenceau, the great painter was trying to 'expose' in his garden at Giverny (Holmes 2001: 72).

We can, for our purposes, cut through such debates by recalling, first, that nature is being understood as natural environments relatively free from human artifice and, second, that the issue is not whether or in what sense the garden may be natural, but whether appreciation of the garden is assimilable to that of nature in the relevant sense. It is perfectly possible to hold that a landscape garden—'Capability' Brown's Blenheim, say—is not natural in the relevant sense, but that appreciation of it may be of the same kind we extend to places that are natural. (Some visitors to Blenheim, one suspects, don't even realize that the park and lake were produced by Brown's workmen, not by Mother Nature.)

What reasons, though, might be given for thinking that gardens more generally, including those which bear artifice on their sleeve, are appreciated in the manner that nature is? Two reasons, ironically, invoke considerations that, earlier in the chapter, were deployed in support of the assimilation of garden to art appreciation. Take, first, the point that many gardens—those of China and Japan, especially— 'imitate' or otherwise represent natural scenery. For the authors cited on pp. 23–4, this meant that such gardens were appreciated for their representational skill and artistry, rather as a portrait is admired irrespective of the qualities of the person who sat for it. Other writers, however, draw a different lesson: in appreciating a garden, we are, in a dog-legged way, brought to appreciation of the landscape or scenery it 'instantiates'. The Japanese garden, it has been said, works through 'evoking … the same feelings one had when actually viewing them [trees, mountains, lakes or whatever] in nature' (Slawson 1991: 58). Whatever it is that accounts for admiration of the Amanohashidate sandbar, this also accounts for our admiration of the garden of the Katsura Imperial villa that 'renders' this geographical feature.

Take, second, the point that gardens are 'nature transformed', 'improved', or 'methodised'. If one stresses the past participles, the idea is liable to emerge of gardens appreciated as art—as transforming, improving, methodizing activity and craft. Stress the noun, however, and the thought becomes that, for all such intervening craft, it is essentially *nature* that we are confronted with in a garden. This, perhaps, was Horace Walpole's point when referring to gardens as nature 'polished': for by polishing—by removing brambles from an oak-tree, say—we 'restore' to nature its 'honours' (in Hunt and Willis 1988: 316). If art must 'set foot in the province of nature', warned another eighteenth-century figure, William Shenstone, this should be 'clandestinely and by night' (p. 293)—otherwise it detracts from the proper object of admiration, nature itself.

A final line of thought redeploys some objections to the art model of garden appreciation in the service of a positive case for the nature model. Natural places engage, as do gardens, all of the senses. More importantly, those aspects of the physical relationship between viewer and viewed which distinguished garden appreciation from that of paintings are also striking aspects, so it is argued, in the case of viewing natural scenery. The viewer is *in* and surrounded by the natural landscape being admired; is typically *moving* through it, and hence is *active*; the landscape is not *framed* for the viewer, and nor are there *privileged viewpoints* from which to regard it. These affinities between the experiences of gardens and of natural places imply, it is claimed, that the appropriate approach to garden appreciation is not a 'traditional' aesthetics of disinterested contemplation, but an 'aesthetics of engagement' of the kind that Berleant has proposed for articulating our appreciation of nature (see Miller 1998: 277). Gardens, like natural places, are not so much 'objects' of the aesthetic gaze, in the manner in which artworks have 'traditionally' been treated, as 'occasions'—to use Berleant's term—for active, engaged experience.

How persuasive are these grounds for the assimilation of garden to nature appreciation? Not very, in my judgement. But nor is one objection that people may be immediately tempted to raise. This is to the effect that gardens just don't look much like relatively wild, uncultivated natural environments. But if this were the only objection, then it would not apply to those many gardens that actually do look rather like natural landscapes—such as Addison's, which, apparently, an ignorant foreigner 'would look on...as a natural wilderness' (quoted in Howard 1991: 33). Anyway, the issue of resemblance is a red herring. For one thing, not all natural places look natural: there are those that an ignorant foreigner would mistake for man-made ones. But that hardly prevents the knowledgeable local from appreciating them as nature. The main point, though, is that, even if all gardens were unmistakably artificial and all natural places unmistakably not man-made, nothing follows as to whether appreciation of a garden is assimilable to that of uncultivated nature. That the polish is apparent would not affect Walpole's point that it is nevertheless as *nature* polished that the garden is admired; nor would it damage the contention that, as with natural landscapes, supposedly, we 'engage' with gardens as 'occasions', rather than view them as art 'objects'.

There are more serious objections to the nature model, however. To begin with, just as it is possible, as we saw on p. 27, to exaggerate similarities between experiences of gardens and of paintings (or poetry), so it is possible to exaggerate how different some garden experiences are from ones of artworks. Jekyll's metaphor of 'garden-pictures' has, we noted, its dangers: but it cannot be denied that sometimes a person just stops and stares at a flower-bed, admiring it in a way that is comparable to enjoyment of a combination of colours on a canvas. We are not always 'engaged' with the garden—moving through it, digging it, crushing leaves for their scent and the sound of their crinkling, and so on. Sometimes the garden, or some bit of it, is the 'object' of a quiet, attentive, even contemplative gaze, not a theatre providing an 'occasion' for experience. (Actually, the same is also true of nature, which is not always experienced in the muscular, 'engaged' manner of the backpacker or kayaker. Conceding this point, however, won't help the 'engagement' theory, since, on that approach, it is the 'engaged' mode of nature appreciation—by way of contrast with the 'disinterested' one 'traditionally' associated with art appreciation—that provides the model for garden appreciation. (See Cooper 2005).)

A second objection derives from an insight hinted at by Kant, and later articulated by Theodor Adorno (1997: 70) when referring to an 'essential indeterminateness' in the appreciation of nature. The point is clearly expressed by Malcolm Budd: 'there are', he writes, 'no constraints imposed on the manner of appreciation [of nature] ... that parallel the constraints imposed by the categories of art'. There is a 'freedom integral to the aesthetic appreciation of nature ... which means that much more is up to the aesthetic observer of nature than of art' (Budd 2002: 147 f.). There are at least three dimensions to this freedom or indeterminateness. Consider, for example, the appreciation of mountains. There are no privileged viewpoints or contexts, as there typically are in the case of artworks, from or in which to experience them. A mountain is equally appropriately viewed from its summit, from an aeroplane above it, from the valley below, from a ledge half-way up it, and so on—and at any season and any time of day (or even night). Next, it makes no sense, in the case of a mountain, to apply to it the types of categories which, in the case of artworks, constrain how they are appropriately to be responded to and appraised. A mountain does not, for example, belong to a genre or a tradition, and has no intention behind it. Finally, mountains do not have functions in the sense that artworks typically do, be they to express feelings, present 'significant forms' to the viewer, 'make a statement', or whatever. Hence it cannot be part of the appreciation of mountains to judge their success or failure in performing a function.

In each of these respects, gardens differ from mountains, and in some of them, at least, stand closer to artworks. Appreciation of them does not, therefore, enjoy the same radical freedom and indeterminateness of nature appreciation. While there is no privileged or authoritative viewpoint or context for experiencing a garden, there are certainly limits—constraints—on appropriateness of experience. One cannot appropriately appreciate a modest-sized garden from a hot-air balloon, for example, or a luxuriant tropical garden, designed to be enjoyed for ten months of the year, during the leaden, rainy days of the remaining two months. Again, there are genres and traditions of gardening, as there are with other art-forms, and gardens are normally designed with a number of intentions in mind. Finally, gardens, like artworks, have functions, including ones of a 'practical' or 'utilitarian' type—growing food, enabling privacy, and the like—that mountains do not have. For such reasons, one may speak of gardens as succeeding or failing. Bouvard and Pécuchet's garden—with its tomb placed in a spinach patch, its Venetian bridge over the runner beans, its yew-trees shaped like stags or armchairs, and its randomly placed sunflowers—was 'quite dreadful' (Flaubert 1976: 61).

These differences, in degrees of freedom and indeterminateness, all stem, of course, from the fact that gardens, unlike mountains, are human artefacts imbued with purpose. And that prompts a final, related objection to the assimilation of garden to nature appreciation. One way of summarizing the previous objection would be this: someone who honours no constraints—those of appropriate view-point and those imposed by categories and functions—when experiencing a garden is not appreciating it *as* a garden, or at any rate not as the garden it is (a kitchen garden, say, or a tropical, 'wild', or tea garden). I might well, from my hot-air balloon, enjoy seeing that little, sunlit spot of green below, breaking up the grey monotony of a townscape: but I am not enjoying it as a garden. Now it is equally crucial to distinguish between enjoying some natural place *tout court* and appreciating it *as* part of nature. The two need not be the same. Someone who enjoys the landscape of Provence only for its associations with the paintings of Cézanne and Van Gogh is not enjoying it as nature. A person who mistakenly believes that a natural meadow is a Robinson or Jekyll creation, and admires it as such, is not admiring the meadow as nature. Conversely, someone who thinks that the park at Blenheim is a natural landscape may be enjoying it as nature, albeit under a delusion.

The question arises, therefore, of what it is to appreciate nature *as* nature. Whatever the full answer to that question, part of it is surely clear enough. Natural landscapes, though they may not be 'pristine' or 'wild', are ones that are relatively unaffected by human artifice.

(Doubtless this needs tightening up. How unaffected is 'relatively unaffected'? Is a tree- and grass-clad hill of the future that is an ex-landfill site of today, left alone for 200 years, a natural place? The point, then, is not precise: but that doesn't prevent it being clear enough for present purposes.) This answer has an immediate consequence: if to appreciate a garden is to appreciate it as a place transformed by human artifice, then its appreciation is not of something as nature. It makes no difference if the garden is a very 'informal', 'natural-looking' one: to enjoy it as nature would still require one either to be ignorant that it is a product of artifice or somehow to put out of mind that it is. So appreciation of a place as a garden cannot be assimilated to appreciation of a place as nature.

Someone may respond that it is a tautology that people cannot enjoy a place both as unaffected by humanity and as transformed by human artifice—and nothing of moment can turn on a tautology. But it is essential to recognize the importance of what is encapsulated in the little phrases '*as* nature' and '*as* a garden'. Doubtless it can happen that a person looking at and enjoying an array of bushes and flowers may be blithely indifferent as to whether the array is natural or designed. But that is not the norm. As Malcolm Budd explains, normally an experience 'under one description [e.g., 'garden', 'transformed by artifice'] has a different phenomenology from that of an experience under an incompatible description [e.g., 'natural place', 'unaffected by humanity']' (Budd 2002: 12). The truth in this is confirmed by the so-called fakery test, which indicates that it is typically integral to the appreciation of natural places that they are taken to be natural, not artefacts. Explain to someone that the forest he has been admiring is composed of trees made from a new plastic that replicates the look, smell, and touch of the real things, and it is probable that his admiration will either evaporate or modulate into a quite different admiration—for the ingenuity of the scientists, say. A reverse fakery test would indicate that it is integral to the normal appreciation of gardens that they are recognized as human transformations of natural places. Explain to someone that the Japanese tea-garden he thought he was admiring is actually a fenced-in area otherwise untouched by human hand, replete with bamboos and stones that just happen to be spaced as if placed for walking on, and the admiration will either evaporate or modulate into a quite different one—for 'the miracles of nature', or whatever.

The import of this reflection might be put in terms of significance. More today, perhaps, than ever before, it is intrinsic to the significance and pleasure people experience when communing with nature that they are aware of the relative freedom from human artifice of the places they are in. Whatever it is that profoundly matters to people receptive to such experience, it cannot be what matters most in the experience of gardens. Only the viewer of a garden who does not know that this is what he is viewing could be moved by a sense of freedom from artifice.

With the case against assimilating or reducing garden appreciation to nature appreciation, it follows that neither of the prime candidates, art and nature, for serving as the model for garden appreciation is acceptable. They fail for symmetrical reasons. The art model insufficiently heeds the fact that gardens are transformations of natural places, containing natural things and subject to natural processes, while the nature model insufficiently heeds the fact that gardens are the products of *human artifice*. With the failures thus diagnosed, a remedy will suggest itself to many people—an operation, in effect, that saves the truth in each model, cuts away what is wrong in them, and splices them together. The garden is neither art nor nature: it is art-and-nature. This is the view examined in the next chapter.

3 ART-AND-NATURE?

Factorizing

A common Japanese name for garden, *teien*, derives from two terms meaning 'wildness' and 'control'—'nature' and 'art', one might almost say (Keane 1996: 14). The name, it seems, encapsulates a persistent and, some might say, obviously correct thought, to the effect that garden appreciation is a combination of art and nature appreciation. This is a thought, moreover, towards which we saw the symmetrical criticisms of the two assimilationist models of the previous chapter apparently tending. If the garden isn't either art or nature, that's because it's both—art-and-nature.

It would be impossible to deny a truth to the claim that appreciation of a garden may be directed towards both artistic and natural aspects. A garden, after all, contains living, natural things that may be admired as such. At the same time, artistry has been invested into the making of many gardens, and this too may be admired as such. But these truisms fall well short of constituting what I shall call the 'factorizing' thesis, one that purports to provide a coherent and comprehensive understanding of garden appreciation. On that thesis, garden appreciation may be analysed in terms of, or factorized into, two modes of appreciation—those of artworks and nature respectively. Put differently, it is the joining together of those two modes, and their application in tandem, that yields a satisfying account of our appreciation of a garden. (If this sounds vague and metaphorical, that is because it is. As we shall see in the following section, a major difficulty for factorizers is so to state their thesis that it is neither boringly truistic nor demonstrably false.)

The factorizing thesis will be familiar to readers of countless popular writings on gardens in which talk of 'combining', 'fusing', 'integrating', or 'uniting' art and nature is pervasive. It is, I take it, the thesis intended by one historian when referring to the idea of the garden as a 'subtle integration of nature and art so that the final result... seems to partake equally of both... a fusion' (Andrews 1999: 55). Or by one philosopher when he writes that in the case of 'non-pristine nature', including gardens, 'appreciation... requires two forms of aesthetic appreciation to function hand in hand... a mixture of the aesthetic appreciation of nature as nature with an additional element... based on human design or purpose' (Budd 2002: 7).

As with some of the claims examined in Chapter 2, it is important not to confuse the factorizing thesis with proposals that might be similarly worded—and important, too, not to contrast it with claims that only sound to be in opposition. When, for example, the novelist and authority on Italian gardens, Edith Wharton, writes that the Italian garden shows how 'nature and art might be fused', she is proposing, *not* that the garden itself is such a fusion, but that it succeeds in relating the house to the 'enclosing landscape' (Wharton 1988: 7). Again, when Denis Diderot ascribed the success of the garden at Marly to 'the contrast between the delicacy of art in the bowers... and the rudeness of nature in the dense bank of trees' (quoted in Adams 1991: 127), he was not rejecting the factorizing thesis. After all, one way in which appreciation of certain gardens might be said to function is through combining enjoyment of starkly contrasting elements, between which, as Diderot puts it, there is 'continual transition'.

It is worth noting, too, that the factorizing thesis is not the monopoly of garden lovers. Indeed, for some authors, it is precisely because the garden is, as they see it, an uneasy combination or fragile fusion of art and nature that it is not something to admire. This, in effect, was Hegel's verdict, discussed in Chapter 1 in the context of explaining philosophy's relative neglect of the garden. Gardening is an 'imperfect art', since its products, being so much affected by nature, are not those of man alone: at the same time, nature as we find it in the garden lacks the 'greatness and freedom' it enjoys when left alone by man (Hegel 1975: 699). Garden appreciation, then, indeed factors out into art and nature appreciation, but in degraded forms of both those two modes.

Still, factorizing does not generally or necessarily require a deprecatory attitude towards gardens like Hegel's. No such attitude could be attributed, for example, to the nineteenth-century American landscape designer and author A. J. Downing. Our enjoyment of some gardens, he writes, owes to the 'polish [that] art can bestow'; that of others to admiration of what is intimated of 'raw', 'rude' nature. But the gardener has 'reached the ultimatum of his art', when his or her creation enables a combined appreciation of what art has bestowed and what has been 'half-disclosed' of nature (Downing 1991: 51 ff.).

Although the factorizing thesis need not involve deprecation of the garden, it does nevertheless have a significant deflationary implication. If the thesis is right, then there is nothing *distinctive* about garden appreciation, for this turns out to be, simply, the combination of two modes of appreciation already and independently exercised, and with their own objects of appreciation, artworks and natural things. Gardens, for the factorizer, are merely particular objects or 'occasions' that invite already established modes of appreciation. A world without gardens would not be one in which a kind of appreciation is lost, but merely one that is devoid of certain objects or 'occasions' for the joint exercise of familiar kinds. To ape Nietzsche's famous remark on music, a world without gardens would be a mistake—but not a very big one, for artworks and natural places would still be there to invite and engage our modes of appreciation.

This deflationary implication must surely be unwelcome to many garden lovers and people for whom gardens matter seriously. The factorizing thesis, one might say, cannot honour what Tim Richardson (2005: 1) has called 'the *unique* strength' that the medium of gardening seems to possess. It is one thing, however, to find a thesis unwelcome, another to show that it is wrong. I won't turn to direct criticism until the next section, but it will be useful, in the remainder of this one, to counter the impression, given by much of the popular literature, that the thesis *must* be true.

First, we should remind ourselves not to confuse the thesis—a substantial and contentious analysis of garden appreciation—with various truisms. That some garden making is an art, and that gardens typically contain natural things, do not entail that appreciation of a garden factorizes into two different modes of appreciation. Second, we should resist the following shoulder-shrugging response to the recognition that gardens are transformations of natural places by art—'What *else*, then, can garden appreciation be except an amalgam of art and nature appreciation?'

One line of resistance is suggested by the implausibility of analogous responses in the case of other, 'mixed' mediums of creative activity. Consider ballet, typically or perhaps truistically a combination of movement and music. So what else, someone may respond, can enjoyment of a ballet be but a combination of enjoyment of the music and enjoyment of the dancers' movements? But that's a bad response. A ballet may be enjoyable even though the music, as listened to on a CD, isn't; nor are the movements when watched by someone with ear-plugs. Appreciation of a ballet, that is, is poorly analysed as appreciation of a piece of music and appreciation of movement which are then stuck together. (Compare and contrast the admiration of the judges, at the Olympic games, for a diver's performance. Like the marks the judges give, this is supposed to be divisible into two components: admiration for 'technique' and 'style', respectively.)

A historical observation on the ballet and music example is pertinent here. Even in cultures where a taste for 'pure' or 'absolute' music—designed simply to be listened to—has developed, this has often been a late development. In many cultures, that is, over a long period, music has been played and heard only as an ingredient in, or accompaniment to, certain activities—ceremony, for example, or dance. It would clearly be problematic, in trying to explain the appreciation of such activities in these cultures, to factorize it into, *inter alia*, musical appreciation: for there would not have existed a tradition of appreciating music apart from its role in these activities.

Now an analogous observation, resulting in a further reason to resist the 'What else?' response, may be made concerning gardens.

'Natural beauty', remarked Adorno (1997: 65), 'is at its core historical', and a great deal of historical research, by Keith Thomas (1984) and others, confirms Adorno's impression that, at least in the post-classical West, a tradition of nature appreciation is of only recent vintage. We know, for example, that the mountains and deserts that today's tourists flock to see were, until the eighteenth century or even later, a rare aesthetic taste. More crucially, we also know that in medieval and Renaissance times there was already appreciative enthusiasm for gardens. But in that case, and however things may stand with us in the twenty-first century, it is hard to see how the garden appreciation of people in those times could have been a partial function of their appreciation of nature. If, as a historian of the experience of mountains asserts, no 'widespread appreciation of views was in operation in Europe before the eighteenth century' (Macfarlane 2004: 145), then appreciation of gardens up until then can hardly be explained, even in part, as the application to gardens of perceptions of nature. This could not, for example, be the explanation of the fondness for gardens of Daniel Defoe, who found 'wilderness' to be 'horrid and frightful to look on . . . good for nothing' (quoted in Adams 1991: 141).

Parallel remarks might be made about traditions of garden appreciation ante-dating, or at any rate not preceded by, traditions of art appreciation. In the ancient empires of Persia, China, and Japan, it seems that a sentiment and taste for gardens were at least as early as those for paintings or sculpture. In that case, and however things stand with us today, garden appreciation cannot always have been a partial function of already established traditions of art appreciation.

It is worth contrasting gardens, in the light of these remarks, with environmental artworks. The 'earthworks' and other creations mentioned in Chapter 2 can only be properly appreciated, arguably, by people like ourselves who are heirs to independent traditions of art and nature appreciation. After all, the creators of such works self-consciously invoke and comment on—perhaps attempt to 'deconstruct' the boundaries between—those traditions. Maybe some modern garden designers do the same, so that someone ignorant of these traditions is not in a position to understand what they are attempting. But it cannot be a general, let alone a necessary or self-evident, truth that garden appreciation breaks down into art and nature appreciation if there have been cultures in which one or other of those factors has been missing. One or other or, perhaps, both: for there may be cultures in which, though artworks and natural places have been appreciated, no distinction between these as objects of two different modes of appreciation has been salient for members of the cultures. Maybe, for instance, Allen Carlson is right to speculate that Japanese aesthetic appreciation 'presupposes a unity of... the artificial and the natural, and not the separation... that characterizes [modern] western aesthetic appreciation' (Carlson 2000: 173).

For the above reasons, then, we should resist the 'What else?' response to the fact that gardens are natural places transformed by human artifice. The assimilation of garden appreciation to a 'mixture' composed of two different factors is a contentious proposal, not a trivial or obvious one guaranteed by the banality that, in some way, artistry and nature are both involved in the making of gardens.

Phenomenology and 'Atmosphere'

It won't have escaped notice that the factorizing thesis has been couched so far in figurative terms: art and nature appreciation are 'mixed' or 'fused' or 'factored together' in garden appreciation. It is reasonable to request proponents of the thesis to articulate it in a less figurative way: otherwise no one is in a position to assess it. Unfortunately, few of them have responded to this request. Nevertheless, it's possible to discern, among some proponents, how they intend the thesis to be interpreted. I shall argue in this section that the thesis, on that interpretation, is implausible. (In the next section, I argue against the thesis on *any* interpretation that I am able to imagine.)

What factorizers seem to have in mind is that our experiences in and of a garden oscillate between those which attend, respectively, to its artistic and natural features. Appreciation of gardens is then held to be a function of those attentive experiences. As one philosopher, reflecting on appreciation of various places—ruins and 'earthworks', as well as gardens—that do 'not fit comfortably' into either the natural world or the world of art alone puts it, 'our perceptual consciousness shifts back and forth between an awareness' of their human, artefactual aspects and consciousness of natural phenomena, such as destructive 'forces of nature' in the case of ruins (Crawford 1983: 55). The factorizer may go on, as indeed the author just cited does, to elaborate this simple proposal in various ways. He might, for example, postulate a third type of awareness—that of the relationships (contrast, transition, or whatever) between the artefactual and natural aspects between which consciousness 'shifts'. Or maybe he will postulate a 'second-order' awareness of the 'first-order' types of perceptual consciousness already mentioned. (The garden appreciator is, in part, appreciating his own appreciative experiences.) To anyone sympathetic to the factorizing approach, such elaborations will doubtless be welcome. But the criticisms I shall make are not of a sort to be allayed by them. They are icing on a cake that will already have crumbled.

In the next section, I try to show that appreciation of a garden *as* a garden could not be explained in the terms of factorizing. For the present, however, my ambition is more modest: to show, on phenomenological grounds, that factorizing, as just interpreted, cannot accommodate a crucial dimension of our appreciative experience of gardens. It cannot account for experience of what I shall call 'atmosphere'.

The word 'atmosphere' is one of those, alongside 'mood' and 'feeling-tone', used to render a term prominent in Japanese garden literature, *fuzei*—which may refer, for example, to that nevertheless enjoyable sense of melancholic pathos (Japanese *aware*) which some gardens induce (Slawson 1991: 137). Nor, of course, is talk of a garden's atmosphere unfamiliar in Western garden literature. 'What matters', writes the garden designer and author Mary Keen (2001: 112), is not the plants, with which too many gardeners are preoccupied, but 'atmosphere', that 'elusive feeling' that she herself tries to capture through her designs. In this connection, one thinks, too, of the pervasive insistence by garden writers on achieving *unity* in the garden—a discernible and satisfying quality of the garden as a whole (see, e.g., Crowe 1994 and Page 1995).

While I want the term 'atmosphere' to retain the affective charge—the dimension of 'mood' and 'feeling-tone'—invested in it, I shall deploy it more in the manner of certain phenomenologists, notably Maurice Merleau-Ponty, for whom it is virtually equivalent to what they call a 'field of presence'. Although it is generally true that, when in 'communication' with an atmosphere or field of presence, our experience is affectively charged—'mooded' or 'feeling-toned', as it were—this is not the only, or even the most crucial aspect, of atmosphere for these authors.

For Merleau-Ponty, the atmospheres of places are entirely familiar to us, yet rarely remarked upon and discussed—partly because of their utter familiarity, partly because they are often hard to articulate, and partly because, once we pay close attention to what we are experiencing, it is no longer an atmosphere, but particular objects and properties, on which we tend to focus. It would be a mistake, however, to suppose that our initial experience of a place takes the form of perceptions directed towards particular objects. An 'explicit perception of a thing'—the 'concretion', in effect, of a 'setting'—presupposes 'a previous communication with a certain atmosphere', a 'field of presence' into which we are 'introduced' (Merleau-Ponty 2002: 374). Merleau-Ponty illustrates the point by recalling his first arrival in Paris. Walking through the city from the station, his experience does not consist of 'perceptions'—that is, of 'positings' of objects as ones of such-and-such sorts, or as standing in spatial and other relations to one another. Rather, particular faces, cafés, and trees are noted only through 'stand[ing] out against the city's whole being' and against a backdrop of 'a certain style or … significance which Paris possesses' for him (pp. 327 f.). Only upon a field of 'latent significance' presented by the city do 'perceptions emerge as explicit acts' (p. 328).

The general point here, which is as applicable to landscapes and gardens as to cities, is fairly clear, I hope. Our initial experience of a place, typically, is not directed towards particular things and properties, but rather registers a style or an atmosphere, a sense of the place as a whole that precedes attention to its constituents. Our first experiences of places rarely take the form of just staring at them from a distance: we move through them, and are otherwise engaged with them. It is not, of course, that I am unconscious of the trees, paths, and so on: asked what I could see, I would unhesitatingly reply that there were trees, paths, or whatever before or around me. But initially, it is not these that are salient: they are not, in Merleau-Ponty's terms, 'posited' and made objects of 'explicit acts' of attentive perception. Typically, too, my experience of atmosphere—the way in which a field of presence communicates itself to me—will be 'mooded', affectively charged: the style of the place is sensed as spooky, cheerful, peaceful, threatening, or whatever. But even if it isn't, it is still an atmosphere—bland and unremarkable, perhaps—that I am initially in communication with.

This strikes me as a compelling account of the phenomenology of our ordinary experience of places. But how, exactly, does it damage the factorizing thesis? First, let's remind ourselves how important a sense of atmosphere is in the appreciation of gardens, indeed of places in general—cities, houses, mountain terrains, and so on. A sense of a garden's atmosphere—of how it 'presences' as a whole—is not some mere initial response that is quickly replaced by the more serious mode of reflective, analytical attention to particular features. Nor is it the fleeting, trivial by-product—a mere 'sensation'—of other modes of attention; something briefly to note in one's diary, perhaps, but nothing to waste words on in a literary appreciation of the garden. On the contrary, it is surely often the case, when we recall and reflect upon our experience of a garden, that 'what we conjure up is not a forensically formulated table of all the variables that worked on us...but a single overall feeling of the essence of the place' (Richardson 2005: 4 f.). Nor is the point restricted to recollections of gardens we have only visited, perhaps briefly, on an occasion.

It applies to ones with which we are very familiar, notably our own. When, during nostalgic moments abroad, I think of my own garden, I find that what I 'conjure up' is indeed its atmosphere or 'presence', not a forensic table of its parterres, gravel paths, or fruit trees.

The failure, then, to accommodate a sense of atmosphere is a serious one in any thesis that purports to account for the appreciation of gardens. And the factorizing thesis does fail. The reason for this is a consideration that was already emerging in Merleau-Ponty's discussion, but which it is worth making more explicit. What that discussion indicated was not only the importance of, but the *primacy* of, atmosphere. What is intended by this term is, first, a chronological point. Typically, as with Merleau-Ponty's first encounter or 'communication' with Paris, it is an experience of the atmosphere, 'style or significance', of the place that precedes explicit acts of attention directed towards particular things and features. (Rather as, to use Merleau-Ponty's own analogy, it is the whole demeanour or 'essence' of a person whom we encounter for the first time—the atmosphere of his or her body, one might say—that strikes us, not the colour of the eyes, the shape of the ears, or the speed of the hand movements in particular.)

But there is a more important point to speaking of the primacy of atmosphere. It is not simply that the explicit acts of perception come after the initial experience of atmosphere. In addition, they are crucially shaped and guided by it. This or that object of a subsequent perceptual act 'stands out' and 'emerges' as an appropriate object of explicit attention only in the context of the whole 'setting' that our introduction to an atmosphere or field of presence provides. As Merleau-Ponty puts it, perceptions of the objects are 'extracted from' and 'presuppose' an experience of the field as a whole.

If this is so, then any attempt to analyse or factorize the experience of a garden's atmosphere into discrete acts of awareness—of artefactual features and natural aspects included—must fail. The main consideration is not that these acts occur after the initial experience of atmosphere, but that which acts are performed are shaped, even dictated, by the latter. Why, for example, should a person's attention turn to a path as it disappears behind a yew hedge, rather than to the hedge itself or to a rose bush close by, unless he or she is already struck by an atmosphere of mystery or invitation attaching to the garden? Why should it home in on a particular gnarled old tree by the edge of a lawn unless the atmosphere that 'presences' is that of a place steeped in history, imbued with significant memories?

There is a response to this line of thought that needs addressing. It will be said that the atmosphere—the 'style and significance'—of a garden must be due to *something* about the garden. It is not some free-floating aura that, like the dew, just happens to settle on this or that garden. Now what can this 'something' be except the things that constitute it, their organization and their various properties? Hence the experience of the garden's atmosphere *must* be due to awareness, however dim and rapid, of these constituents, and so to particular acts of perception divisible into those directed at artefactual and natural constituents respectively.

There are two things wrong with this response. First, we need not grant that things, structures, and properties to which the garden owes its atmosphere are available to conscious perception and attention. The great American landscape architect and pioneer of the National Park system, Frederick Law Olmsted, wrote:

Beauty, grandeur, impressiveness [of] scenery, is not often to be found in a few prominent, distinguishable features, but in the manner and the unobserved materials with which these are connected and combined. Clouds, lights, states of the atmosphere, and circumstances that we cannot always detect, affect all landscapes. (Olmsted 1852: 154)

He is surely right. A garden may owe its 'mood' to features some of which are not open to the viewer's unaided perception and others of which, though not invisible to him, could only be explicitly recognized for what they are or for their likely effects by an expert, a professional landscape designer, perhaps, or a psychologist. It is such features—the exact orientation of a house and garden, for example, or the direction of air-flow—that *fengshui* designers exploit in producing a place's effects upon the people who live in it.

But even if all the features, artefactual and natural, contributing to the atmosphere of a garden were available to perceptual awareness, it would be wrong to conclude that appreciation of the atmosphere was due to awareness of them. As Merleau-Ponty, once more, argues, the fact that my 'total perception'—of an atmosphere or field of presence—is 'capable of dissolving' into 'analytical perceptions' of particular features does not mean that it is 'compounded of such analytical perceptions' (2002: 342). The mistake, to return to his analogy, is akin to supposing that, just because I can come to focus on a person's eyes, ears, and hands, my experience of his or her whole demeanour or presence—the atmosphere of his or her body—must have been 'compounded' of such perceptions. And this is so, even if my experience would have been different had the person's eyes, ears, or whatever themselves been different from what they are. That a garden would not have the atmosphere it has without the natural vegetation it contains and without its particular design features is no reason for holding that appreciation of this atmosphere is a function of 'shifting' awareness of these factors.

The factorizing thesis, as we have been interpreting it, is a claim about how the appreciation of a garden is 'compounded' of conscious experiences of its various aspects. The phenomenology of appreciation of a garden's atmosphere shows, however, that, even if this appreciation may 'dissolve' into acts of awareness of these aspects—in the sense that a person may come explicitly to attend to them—it is typically not 'compounded' of them. But does the factorizing thesis have to be interpreted in the above manner—in terms, that is, of 'shifts' between awareness of artefactual and natural features respectively? Perhaps not, but I now turn to a criticism of the factorizing approach which will militate against it on any interpretation.

Relevance, Identity, and Holism

So far I have treated references to art and nature as 'fused', 'mixed', 'integrated', 'combined', and so on in the garden as equivalent figurative expressions of the factorizing thesis. Perhaps that is wrong: perhaps a metaphor like that of fusion, especially when accompanied by talk of 'dialectical' or 'dynamic' relationships between art and nature, gestures towards a different thesis. Thus, when two authors speak of some topiary as failing 'to work dynamically with nature's forms', their point is the normative one that such topiary 'constitutes the *imposition* of a human view onto nature' (Brook and Brady 2003: 135). And perhaps Donald Crawford's (1983) talk of a 'dialectical' relationship between artefact and nature in gardens should be construed, not as I construed his remarks on p. 48—as a statement of the factorizing thesis—but as gesturing towards a view, like my own, critical of it. Maybe 'dialectical' indicates a relationship too intimate to be broken down into separable 'factors' that are subsequently 'combined'. (Crawford, after all, speaks of trying to identify a relationship between elements that are 'interacting in other than purely harmonious and straightforwardly causal ways' (1983: 49).)

Be that as it may, I now develop an argument against factorizing that is effective against the thesis on any reasonable interpretation of it. While the argument draws on the previous section, the charge is no longer one of failure to provide a plausible phenomenology of an important dimension of garden appreciation. The charge, now, is the more radical one that any factorizing explanation of garden appreciation is circular, and hence not a genuine explanation at all. More fully: the factorizing approach requires that appreciation of a garden as a garden results from appreciation of components that have the identity they do—are what they are—independently of their relation to the garden as a whole. But that requirement cannot be satisfied. Appreciation of a garden as a garden cannot be thus reduced, since the features relevant to its appreciation cannot be identified independently of their relation to the whole garden. Their relation to the garden is, as logicians would say, an 'internal' one, without which they are not the features or components they are. Care is needed here when talking of 'the garden as a whole'. As noted in Chapter 1, we often speak of what, at one level of classification, is a single garden, such as Mottisfont Abbey Garden, as containing, at another level, several gardens, such as the walled rose garden at Mottisfont. So, when claiming that a feature is 'internally' related to the garden as a whole, we need to decide what, in the context, counts as this whole—the rose garden alone, say, or this together with the octagon of yews, the alley of lime-trees, and so on which are collectively referred to as Mottisfont Abbey Garden. Maybe— though only maybe—a certain feature owes its identity to its place in the rose garden, irrespective of its relationship to the rest of the Mottisfont complex.

To develop my argument, let me take up first my mention, in the previous paragraph, of features of a garden that are *relevant* to its appreciation. Not all the countless features of a garden can be relevant to its appreciation—not, for example, the worm burrowing through the soil six inches beneath one's feet. The question arises, therefore, how features relevant to an appreciation, be they artefactual or natural, are identified? Do they just present themselves and

Plate 3. Sculpture and Italian garden, Villa Taranto, Lake Maggiore, Italy

inevitably stand out in relief for the garden visitor who looks at, sniffs, and listens to what is before or around him? The answer is 'No'. Notice, first, that a feature which, in one garden, is relevant to a certain appreciation may, in another garden, either be irrelevant, or relevant to a very different appreciation. The slim silver birch which 'lends height' so effectively in the Smiths' garden may, when dug up and replanted among a group of birches in the Jones's garden, not even get noticed as a distinct feature. The massive statue which, in the Italian renaissance garden, served as an organizing focal point may, when exported to a millionaire's lawn in Florida, 'kill' the previously effective urns and shrubs. The wider point indicated here is that no feature, considered 'in itself', apart from its relation to the garden as a whole, is relevant to appreciation. This point is no longer the earlier phenomenological one, concerning the way in which our noticing this or that feature is guided or shaped by a sense of atmosphere. It is, rather, that whatever gets noticed is only relevant (or irrelevant) to appreciation of the garden in virtue of its relation to the garden as a whole.

This point is liable to be overlooked, because our long experience of, and familiarity with, the types of gardens we generally encounter have primed us in advance, as it were, unhesitatingly to zero in on certain features as relevant. Familiar with the English cottage garden, someone visiting an example of that genre will doubtless, and effortlessly, attend to the clump of hollyhocks, the rose creeping over the trellis, the rustic wheelbarrow by the ivy-clad wall, and so on as significant features that contribute to the effect of the garden. However, we only have to imagine someone visiting a garden of a type with which he is entirely unfamiliar—an avant-garde 'modernist' one, say, or a Mughal *chahar bagh*—to realize that the discernment of relevant features is indeed a function of past experience and familiarity, not a matter of untutored response to something which features wear on their sleeve. I have been in avant-garde gardens where I just don't know if a certain feature—a patch of crudely mown grass, say—is supposed to be relevant to my appreciation of it or not. Someone unfamiliar with Mughal gardens will not only be unable to identify aspects germane to its representational purposes—that of symbolizing 'the four rivers of life', for example—but won't be able to tell whether a certain feature, such as the noise of running water, is a relevant one or an 'accident' to ignore.

Important as this point about relevance is in its own right, it is preparatory to a more radical, 'holistic' point about the identity of garden features and components—to the effect that this depends on the relation of the features to the garden as a whole. The relevance of a feature is tied to its identity: for, generally, it is the feature as experienced—not the feature under some purely physical description—that is relevant to one's appreciation of a garden. What the feature-as-experienced *is*, like the matter of its relevance, therefore depends on this relation to the whole. If so, then appreciation of the garden cannot be compounded from awareness of features independently of their relation to the whole: for these are not the same features as those experienced when appreciating the garden as a whole.

The general point here is nicely made in Wallace Stevens's poem, 'Anecdote of the Jar':

> I placed a jar in Tennessee,
> And round it was, upon a hill.
> It made the slovenly wilderness
> Surround that hill.
> The wilderness rose up to it,
> And sprawled around, no longer wild . . .
> It took dominion everywhere.

In a good sense, Stevens's jar has changed the wilderness (which is 'no longer wild'), just as its location in the wilderness has changed the jar (it now has 'dominion'). Before the jar was placed, the observer saw a 'slovenly', random wilderness; now he experiences it as 'rising up' and 'surrounding' the hill, 'gathered', one might say, about the jar. Despite being physically unchanged, both wilderness and jar are no longer the features, the objects, they were. (This does not contradict a point made on p. 18. There are not two jars, a physical one and a 'virtual' one. There is simply the jar: but we can, and indeed must, distinguish the jar as experienced from the jar under a purely physical description. And it is the former which is relevant to appreciation of the whole Tennessean scene.)

The argument may be elaborated, this time specifically in connection with gardens, by considering experiences of trees. In a sensitive and indeed poetic passage, surely inspired by Heidegger's essay 'The Thing', Roger Scruton writes that 'a tree in a garden is not like a tree in the forest or a field. It is not simply there . . . accidental. It stands and watches . . . converses . . . with those who walk beneath it.' 'Trees standing in a garden', he continues in unmistakably Heideggerian idiom, 'join both earth and sky'; they serve to 'gather' other things in the garden around them, and are essentially participants in that network of 'between-ness [which] is what we see all around us in the garden' (Scruton 2000: 83). Before rendering Scruton's point less poetically, let me cite another literary presentation of it. In Virginia Woolf's short story 'A Summing Up', a woman contrasts her experiences of two trees, one in a garden, the second in a marsh. The tree in front of her in the garden 'became soaked and steeped in her admiration for the people of the house; dripped gold; or stood sentinel erect. It was part of the gallant and carousing company.' The tree in the marsh—for all we know, physically just like the one in the garden—is, on the other hand, 'denuded of its gilt and majesty'. It 'became a [mere] field tree', not 'mated' to anything else (Woolf 1962: 140 f.).

What both passages indicate is that a feature of the garden, a tree, owes its identity to its relationship to the wider whole of the garden: to the ground it stands on, the plants and areas it is 'between', the objects that it 'gathers' around it, and the people who walk or carouse beneath it. Even if it carries the danger of encouraging the notion of there being two trees—a physical thing and a virtual or 'ideal' object—it is surely intelligible and illuminating to say that the tree in Scruton's or Woolf's garden just isn't the same tree as a physically similar one outside the context of the garden, in a marsh, would be. It then makes no sense to suppose, as the factorizer does, that people may first experience and appreciate the tree and then, by compounding this experience with those of other features, come to an appreciation of the garden as a whole. For the tree, like the rest of the garden's features—artefactual and natural—are what they are, as objects of experience relevant to appreciation, only in relation to the whole. There is no need, I think, to spell out the point in connection with the artefactual features of the garden. One could quickly translate Stevens's remarks about the jar on the hill into ones about a jar in the garden and then extend them, without difficulty, to, say, a carefully composed 'drift' of flowers or the shapes of some topiary. Whatever the artistic features in question, they, like the tree, will be the objects of experience they are only through their 'internal' relation to the whole garden.

It is not being denied, of course, that a person may focus on and appreciate a natural feature of the garden in the way in which he or she might do were it a physically indistinguishable object outside the garden. Equally, an artefactual feature of the garden might be experienced and appreciated just as it would be were it outside the garden. But in neither case is anything being appreciated as a feature of the garden, and no compound of such acts of appreciation could amount to appreciation of the garden as a garden. To suppose otherwise would be like thinking that appreciation of a ballet performance could be compounded out of the appreciations of the movements that might be made from the points of view of aficionados of gymnastics and high-jumping.

The case against factorization is now complete, and with it the case against attempts to assimilate or reduce appreciation of gardens to art and nature appreciation, whether respectively or in tandem. There is an important implication to draw from the collapse of assimilationism, but it's worth emphasizing as well what is not being rejected. To begin with, I am not, of course, denying the clichés that have helped to inspire assimilationist ambitions: the typical garden is indeed a natural place, containing natural things, transformed by human artefact, and some gardens, at least, display artistry in application to natural materials. Nor am I rejecting, in the present context at least, a number of claims that might be intended by talk of the garden as a fusion or whatever of art and nature. I am not, for example, challenging the claim made by two authors, already cited on p. 54, that the topiarist should respect 'nature's forms', and not simply 'impose' a 'human view onto nature through aesthetic means'. But nor am I taking issue with the claim, possibly at odds with the previous one, that gardeners should strive to be 'the great humanizers, by which man projects his personality and love of [artistic] creation into the realm of nature' (Crowe 1994: 160). Both claims are normative ones, on which my rejection of assimilationism, which is an explanatory and not a normative position, has no direct impact.

Rejection of assimilationism and factorizing does, however, affect our sense of the importance of gardens. On p. 44, I observed that, if the factorizing thesis is true, then there is nothing significantly 'distinctive' about garden appreciation, and that the garden does not have, as one author put it, 'the *unique* strength' that many imagine. Gardens, on that thesis, are simply certain places or objects on which one may bring to bear types of appreciative competence that are already exercised, and massively so, in the appreciation of other things, artworks and natural places. With the collapse of assimilationism, to recall my parody of Nietzsche's remark about music, it emerges that a world without gardens would be a big mistake, a world that refuses exercise of an appreciative competence uniquely exercisable with respect to gardens themselves.

There is a poem by Reginald Arkell which reads:

> What is a garden?
> Goodness knows!...
> 'Tis just a garden,
> After all.

Here we hear the wisdom of Bishop Butler's maxim that everything is what it is, and not another thing. A garden is what it is, and not something else—art, nature, or art-and-nature. And the same goes for garden appreciation. But perhaps we should take the second line, not as a counsel of despair over any attempt to explain why gardens matter, but as a challenge to continue searching for an explanation. Any explanation must respect the uniqueness of gardens and their appreciation; but unless one wrongly supposes that assimilation is the only proper form of explanation, this should be no deterrent to that search.

4 GARDENS, PEOPLE, AND PRACTICES

'Gardens are for People'

Many pages in the ancient Japanese treatises on gardening—*Sakuteiki* and Zōen's *Illustrations*—are devoted to warning the owners of gardens to respect, for their own well-being, various taboos. We learn, for example, that it is 'exceedingly harmful for a person of the Fire Nature to look at a red rock . . . with a nandina plant facing it' (Zōen 1991: §2). This is early testimony—quaint-sounding, but intelligible in the context of *fengshui* geomancy—to something apt to be forgotten by philosophers for whom reflection on the garden is confined to aesthetics: gardens connect with aspects of human life and well-being that are not confined to that domain. As another early writer, Thomas Hill, reminded his sixteenth-century readers, 'a garden shall workemanly be handled and dressed unto the necessarie use and commoditie of man's life' (quoted in Church 1995: 51). 'Commoditie' may not be the *mot juste* for what I discuss in this chapter, but this is indeed dimensions of 'man's life' that extend beyond the confines of the aesthetic.

In Chapters 2 and 3, I argued against attempts to assimilate the appreciation of gardens to that of art and of nature, separately or in tandem. But even if assimilationist explanations of why gardens matter to us had been more cogent than I allowed, their success, as answers to our 'fundamental question', would still have been limited. At best, they would explain why gardens matter to us aesthetically, and what I want to establish in this and following chapters is that the significance of gardens stretches well beyond what would thus be explained. One way to recognize the limitations of the approaches discussed in the two preceding chapters—of, in effect, an aesthetic focus on the garden—is to note a couple of striking omissions. On none of those approaches, first, do we hear anything of substance said about garden*ing*, that practice in which, we were informed (p. 2), 78 per cent of us participate of a weekend. Perhaps the assumption is that gardening is of importance to people only as a means to creating certain objects for aesthetic appreciation. But that is a rash and implausible assumption: were it true, one wonders why so many people who could afford it do not automatically pay others, or invest in 'hi-tech' equipment, to do the job for them. Second, no significant distinction is made on these approaches between the gardens people merely visit and those which they own, ones which, for a lot of the time, they live in and engage with as part and parcel of their everyday existence. These two omissions are related, for one obvious way in which people engage with their gardens is when they are gardening.

The omissions reflect a more general failure, at least in the philosophical literature, to respect the importance of what I shall call 'garden-practices'. By that expression, I do not mean, simply, such activities as garden designing, planting, digging, and whatever else comes under the heading of 'gardening'—the sort of activities discussed in magazine articles devoted to 'good practice' by gardeners. I mean, as well, a host of activities and practices that are appropriately pursued in the garden, ones for which, one could say, the garden provides an especially hospitable theatre. Recall, for example, some of the pursuits mentioned by Battisti (p. 4) in his catalogue of the functions of the Renaissance garden, which ranged from feasting, trysting, and exercising to philosophizing and botanic research. Or consider Russell Page's observation that the activities which go on in designated areas of the modern garden—play area, barbecue patio, swimming-pool—are not extraneous 'necessities infringing on precious garden space', but themselves 'a form of gardening', at any rate a form of appropriate engagement with the garden (Page 1995: 61). The importance and appropriateness of garden-practices was fully appreciated by the author of the book from which this section takes its title—and the designer of gardens which instantiate his title. In Thomas Church's well-known words, a garden is 'for people', a place to be 'lived in', and should therefore be 'designed primarily for living, as an adjunct to the functions of the house' (Church 1995: 32). (Here is an appropriate place to note a possible discrepancy between UK English and US English uses of the word 'garden'. It has been pointed out to me that some Americans would find talk of swimming or playing in the garden odd: it is the *yard* one swims or plays in. Like Page and Church, I am using 'garden' broadly, so as to encompass what American speakers would call 'yards'.)

I'll shortly address the predictable response that while, yes, some garden-practices are important to people, there's nothing 'interesting'—philosophically, at least—to say about them. First, though, I address another possible critical response. I am contrasting attention to garden-practices with an aesthetic focus on the garden, but it will be said that the contrast is without force in the absence of a definition of the term 'aesthetic', which I haven't provided. Well, nor am I going to, and the contrast doesn't require me to. Unless the term is employed, unhelpfully, to refer to anything we value doing for reasons that are neither 'moral' nor narrowly utilitarian, then such garden-practices as playing on the grass, sharing a barbecued meal with friends, and botanical enquiry are not, even on very liberal interpretations of the term, aesthetic activities.

This does not entail that such practices are necessarily disjoined from aesthetic enjoyment and appreciation. That would only follow on an over-narrow conception—Kant's, perhaps—of 'the aesthetic attitude' which excluded practical concerns altogether. In fact, most of the practical activities we enjoy—while they may not be performed, as Karl Marx hoped all production would one day be, 'according to the laws of beauty'—are engaged in with some attention to sensory appearance and some sense of style. Even the digger of a potato patch in his garden is likely to have an eye for straightness of line and symmetry. Dewey was surely right to bemoan any alleged 'chasm between ordinary and aesthetic experience' and to emphasize 'the continuity of aesthetic experience with normal processes of living' (Dewey 1980: 10 f.). But the fact that a garden-practice is typically 'enhanced', as he puts it, through attention to its aesthetic aspects does not make it one of those, like strolling round the garden to admire its roses, to which the expression 'aesthetic pursuit' would happily refer.

In turning my attention to garden-practices, I am not intending to diminish the significance that aesthetic appreciation of gardens clearly has for many people, still less to denigrate such appreciation as an affectation of the idle and effete. How central to garden enjoyment appreciation of its aesthetic aspects is doubtless varies from person to person. For many people, clearly, the garden's colours, forms, and smells occupy the foreground of their attention; for others not, though, as Dewey's remarks imply, such phenomena, in the case of many garden-practices, may serve to enhance those practices. The garden-pool swimmer may not focus on the branches above or the sounds of nesting birds, but they are there in the background, as it were, to distinguish the swim from one taken at the local leisure centre. Arguably, the centrality or otherwise of aesthetic aspects also varies from age to age, society to society. In his wise and witty *Second Nature*, Michael Pollan observes that 'British gardeners have traditionally regarded themselves more as artists than reformers. The issues in English garden writing are invariably framed in aesthetic, rather than [the] moral, terms' more distinctive of American writing (Pollan 1996: 78). The English garden owner, he is suggesting, is more concerned with creating Jekyll-like 'beautiful pictures' than with, for example, the freedom of movement and opportunity for social gathering that the barer, more open American garden affords.

On a historical note, it is my impression that, in two respects, the garden has become increasingly appreciated in recent years as a theatre of garden-practices, and less as an aesthetic spectacle. In both cases, ironically, recent developments recall much older traditions. First, more and more activities—for economic, technological, and other reasons—are being conducted by more and more people out of doors, in the garden: cooking, dining, swimming, partying, playing music, and so on. Second, both designers and their clients have become increasingly intent, as one influential designer, Geoffrey Bawa, puts it, on 'the breaking down of barriers between interior and exterior' (Robson 2002: 34)—on a 'symbiosis' of house and garden to facilitate a smooth shuttling back and forth between the building and its surroundings as people go about their activities. (Think of the roles played by that especially 'symbiotic' and nowadays ubiquitous structure, the conservatory.) In both these respects, readers of Murasaki's *The Tale of Genji* will be reminded of the Heian Japanese conception of the garden—a place at once the main theatre for the everyday, if unusually refined, activities of Kyotan aristocrats, like poetry readings, and one so minimally separated from the house by sliding paper screens that 'the open, sunny garden was no mere decoration but the very pivot of Heian domestic architecture' (Morris 1979).

Let me now return to the objection that garden-practices, while doubtless important to people, are an unpromising topic for philosophical discussion. Drinking water and taking exercise, it'll be said, are important, too, but no one devotes chapters of philosophical tomes to them. My response to this is the rest of this chapter and the one following. In Chapter 5, I discuss how garden-practices contribute or otherwise relate to 'the good life', a notion—I suggested in Chapter 1 (p. 5)—to which our 'fundamental question' of why gardens matter soon leads. In the present chapter, and partly by way of preparation for that discussion, my aim is not to provide a catalogue or history of garden-practices, but rather to offer something by way of a phenomenology of various garden-practices—illuminating description, I hope, of people's experience of engagement in these practices and of the significance this has for them.

There are two important truths about many garden-practices which I intend my exercise in phenomenology to establish in the following sections. The first is that there is a complexity and interest—a 'depth', even—to these practices that is quite missed if they are described simply as 'enjoyable' or as mere means to certain ends. Thomas Hill was right to remind garden designers to attend to 'the commoditie of man's life', but it would be superficial to hold that the importance to people of even the most practically orientated garden-practices is only of the pragmatic kind suggested by terms like 'commodity'.

The second truth bears comparison with one from the previous chapter. There I argued that the appreciation of gardens is 'distinctive': the objects of appreciation, natural or artefactual, owe their identity to their place in the garden as a whole, which was why garden appreciation could not be reduced to types that are also extended to things other than gardens—to artworks, say, or 'wild' landscapes. In this chapter, I suggest that garden-practices are 'distinctive', and for a somewhat similar reason. These are not practices that just happen to be pursued in the garden, indistinguishable apart from their location from practices pursued elsewhere. It may sound paradoxical to say that swimming or breakfasting in the garden is a different activity from doing so indoors: in either case, after all, it is swimming or breakfasting that someone is doing. But the allegedly paradoxical way of putting it serves to emphasize that a swim or a breakfast in

Plate 4. Swimming-pool and garden, Ta Cenc Hotel, Gozo, Malta

the particular context furnished by a garden may have a 'tone' that makes it a very different event from a swim or a breakfast elsewhere. Our language might have contained more compound expressions, like 'garden-party', than it does—ones which refer to what can only take place in a garden. I'm suggesting, if you like, that we should hear expressions like 'swim in the garden' *as if* they were such compound, hyphenated expressions—by way of an antidote to the tendency to suppose that they refer to certain general activities with the extra, merely incidental feature of taking place in the garden.

The point is an important one if we are to understand how garden-practices bear on the issue of why gardens matter to people. For it might be contended that they have little bearing, that such practices matter, if at all, for reasons having nothing essentially to do with gardens—such as enjoyment of exercise or the benefits of a healthy breakfast. But that contention must be wrong if, as I am urging, it is *swimming-in-the-garden* and other practices inseparable from the context of the garden that are under consideration.

Gardening

'Garden-practice', as explained, does not refer only to gardening, but it is this which the expression would most immediately suggest to a reader. The word 'gardening', in turn, immediately conjures up such more or less muscular activities as digging and pruning—the sort that people would sketch if asked to draw someone gardening— but it also applies, of course, to propagating seeds in the greenhouse and spraying for greenfly. I shall be using the word still more widely—to apply, for example, to ordering bulbs from the catalogues one reads of a winter evening, or sketching out a new fish-pond. Gardening, for my purposes, is just about any activity geared to the design, cultivation, and care of the garden.

It is symptomatic of a primarily aesthetic approach largely to ignore the practice of gardening—the assumption being, I surmised, that there isn't much of interest to say about the practice independently of the objects, gardens, which require it. Gardening, it might be conceded, is a form of self-expression, say; but since, presumably, it is the gardens themselves, rather than the actions involved in their creation, which are the vehicles of expression, then the value of the practice is a function of that of its products. This is a type of assumption, to be sure, that even those adopting a primarily aesthetic focus may want to challenge. Nietzsche, for one, complained that aesthetics was overly obsessed with the 'spectators' of art, to the neglect of its 'creators' (1968: iii. 6), whose motives and procedures are of independent interest. Be that as it may, once we get away from a purely aesthetic approach to gardens, there is no reason to treat the significance of gardening, the practice, as parasitic on that of its products.

Of the 'two truths' with which the preceding section closed—concerning the 'depth' and distinctive 'tone' of garden-practices—it might seem that only the first could be worth stating with respect to gardening. For isn't it self-evident, a matter of definition, that gardening is a 'distinctive' practice in the relevant sense of being essentially related to gardens? One can only garden where there is a garden to garden in. But care is needed here. The names of most component practices that constitute the larger practice of gardening—'digging', 'mowing', 'weeding', and so on—do not, like 'gardening' itself, refer only to what goes on in gardens. People dig graves, mow meadows, and weed pavements. So there is a substantial question to raise—whether or not these practices, when pursued in the garden, have a 'distinctive tone', as I put it; whether, to risk speaking paradoxically, they are different activities from those conducted outside the garden—in the graveyard, meadow, or city street, say. The second of my 'two truths' answers that question affirmatively. Actually the 'two truths' are closely connected. One reason why a garden-practice may be 'distinctive' is that there is a certain complexity and 'depth' to it. Conversely, it may have this complexity and 'depth' because the context of the garden lends it a 'distinctive tone'. Because of this close connection, I won't for the most part treat the two points separately in what follows.

The issue is that of the significance of gardening. Why does gardening matter to people? There are two answers to that question, which I'll label the 'instrumental' and the 'hedonistic', which I want to reject. Or, to be more precise, they are superficial answers that cannot provide complete, universally applicable explanations of why gardening matters. While I shall be gesturing towards more satisfactory answers, I won't in this chapter be aiming at completeness. I don't want to encroach too far on positions I develop in later chapters—on, for example, my discussion of 'the good life' in Chapter 5 and that of 'the meaning of the garden' in Chapters 6 and 7. So my remarks in the present chapter are prolegomena to later discussions.

According to the hedonistic explanation, gardening matters because it is enjoyable and pleasurable: no 'deeper' explanation, therefore, is required. Few gardeners, of course, would deny that gardening has its many and often prolonged pleasurable moments: tending to one's hydrangeas on a balmy summer's evening, feeding grateful mouths in the fish-pond on a spring morning, or turning over sods of Čapek's 'gingerbread' soil, 'light and good as bread'. But few would deny, either, that gardening has its many and often prolonged moments that are anything but enjoyable: carrying pails of water to thirsty flowers during a drought, breaking the ice on a pond during a blizzard, or turning over sods of Čapek's 'sterile matter' of 'primeval clay'. And Gertrude Jekyll, one suspects, was unusual in maintaining that 'weeding is a delightful occupation' (1991: 51).

Perhaps, then, the hedonist's proposal is that a 'hedonic calculus' would show that the pleasurable moments outweigh the rest, by criteria of intensity or duration, so that gardening does produce 'net utility'. But who, one wonders, has ever attempted such a calculation? And if someone who did found that, on balance, the unpleasurable moments prevailed, would he then hang up his spade and trowel and admit to being mistaken in supposing that gardening is an enjoyable practice? Surely not—which suggests that the practice is not explained by the net pleasure which its component activities yield.

The hedonist might complain that his position is being misunderstood: that when he speaks of a gardener enjoying or taking pleasure in his work, he means this in the sense that even a bomb-disposal expert or snake-venom collector may enjoy and take pleasure in their work. What is meant is not the pleasantness of all or most of the activity, but the 'satisfaction' it yields. Now, however, the claim that people garden because it is enjoyable and pleasurable looks unhelpful, for attention must turn to what it is about gardening that makes it a 'satisfying' practice. The hedonist, moreover, can no longer rule out as unnecessary the search for a 'deeper' explanation of the significance of gardening. Indeed, one now wants to question the aptness of his vocabulary, for the suspicion arises that this is being employed vacuously if *every* practice that yields 'satisfaction', of whatever kind, is thereby described as enjoyable and pleasurable. One is reminded, here, of a familiar and deserved criticism of Epicurus. Having announced that pleasure is 'the goal of living', he then develops an account of this goal—in terms, partly, of a 'freedom of the soul' achievable only through philosophical understanding of one's place in the scheme of things—which not only makes this goal sound complex and 'deep', but for which 'pleasure' seems a peculiarly infelicitous label (in Inwood and Gerson 1988: 24). In other words, Epicurus is not the hedonist he at first appears to be.

More plausible, it might seem, is the 'instrumentalist' account of why gardening matters. It does so, on this account, because it is the means whereby gardens are created and kept in being. Why gardens themselves matter—for aesthetic reasons, say, or for further instrumental reasons—is beside the point: whatever the reasons might be, the practice of gardening matters simply because it is a necessary means to there being gardens. I have already indicated one problem for the instrumentalist—to explain why the many garden owners who could afford to do so do not pay others, or install 'hi-tech' machinery, to do much of the work for them. If all that matters is the 'end', the garden itself, it must be indifferent, except on pragmatic grounds of cost, effort, and the like, what means are adopted. This objection prompts the thought that 'caring for the garden is not a chore, but the very point of having a garden in the first place' (Keane 1996: 128). That may be exaggerated, but it reminds us of the important truth that many gardeners know, and *welcome* the knowledge, that in making a garden, they are not creating a finished product, but one whose maintenance, enhancement, and transformation are a long-term commitment, not to a 'chore', but to a rewarding practice. By contrast, almost no one installs windows with the purpose, even in part, of cleaning them: that's why cleaning windows is a 'chore', a mere means to the end of having clean windows.

Alan Titchmarsh, recall (p. 3), spoke of gardening as being, apart from having children, the most rewarding of activities. Whether or not he intended one, there is an analogy—only very partial, of course—between the two. Couples don't have children simply in order to add some new human beings to the world's population; nor do they anticipate raising and caring for their children as 'chores' necessary for keeping those human beings alive, ones which, if there were more money in the bank, could be farmed out to others. Rather, they look forward to their essential participation in the development of their children, creatures who are, moreover, never going to be 'finished products'. Suitably amended, what is true of these couples is true of many gardeners.

In order further to appreciate the limitations of the instrumentalist approach, and to suggest something of the 'depth' of the significance of gardening, let us focus on a species of gardening which might seem to lend itself especially well to instrumental treatment. I refer to kitchen gardening, the growing of food in one's own garden—a vegetable like squash, say, whose purely aesthetic appeal, moreover, is limited. Why do people do this? Ask many of them, and you will likely be told of its 'practical' advantages—it's cheaper than buying squash, the home-grown one tastes better than the shop-bought one, and so on. But one rightly wonders whether all these people would give up growing their own if they were shown that, actually, it isn't cheaper, and if they did badly on 'blindfold tests' of their ability to distinguish home-grown from supermarket squash. And one rightly suspects, in the case of some people, that shyness or difficulty in articulating reasons of a different kind accounts for their offering only 'practical' ones.

One 'deeper' reason—which I shall elaborate on in Chapter 7—why people grow their own is encapsulated in Michael Pollan's remark, apropos a gigantic, ugly Sibley squash he grew, that it was 'a *gift*'. By this remark, he intends at least three things. To begin with, the ripened squash induces in him the sense of wonder and 'magic' that many gifts do, especially in children—one not dispelled by reading scientific tomes on photosynthesis which tell the causal story of the vegetable's life. Second, the squash represents a 'net gain'—a 'free lunch'—in the 'planet's economy': for it is not a lump of redistributed matter but, thanks to photosynthesis, new matter. That this 'newness comes into the world' with one's care and help is, Pollan adds, 'reason enough to garden'. Third, for all the human effort and ingenuity invested in growing the squash, it is nevertheless 'given' to him—a gift of 'grace', in effect, without which no amount of effort would be of use (Pollan 1996: 156 ff.).

I surmise that the experience of home-grown vegetables as 'gifts', in the sense unpacked by Pollan, is shared by many gardeners, even if it is one that they feel it uncomfortable or awkward to articulate. Now, this is not an experience that supermarket vegetables induce in many people, partly, at least, for a reason that is also a more or less independent reason why people grow their own. Let's take our lead, once more, from Pollan, when he remarks that, for his grandfather, 'to work in his garden was to commune with nature' (1996: 18). Those last three words might cause a grimace in readers suffering from a surfeit of popular gardening magazines and 'New Age' books in which they are glibly trotted out, it can seem, on every other page. In the prose of writers like Pollan and Karel Čapek, however, the phrase is no longer glib, but evinces something real and important.

Central to the general idea of communing is that of intimate sharing or mutuality, and this latter idea surely gets a purchase on the relationship between a gardener and the vegetables he or she grows in a way that it does not on that, say, between walkers and the hills or fields through which they walk. To begin with, there is a sharing of interest and fortune: roughly speaking, what is good for the gardener is good for what he grows, and vice versa. After the rain that follows a drought, Čapek writes, 'we have all breathed out: the grass, me, the earth, all of us: and now we are fine' (2003: 78). Second, there is sharing in the form of engagement in a mutual enterprise. It is easy to smile at those who talk encouraging words to their plants, as if these were little people; but when this is viewed, not as a horticultural experiment, but as expressive of a sense of being engaged with the plants in a mutual enterprise—of a shared good— it is no more to be derided than the 'conversations' people have with their pets. But the analogy with the gardener's and the plant's shared good should not, perhaps, be the good that two people may share, but the kind shared, say, by a musician and his instrument in their mutual enterprise of making music. The Bengali poet and composer Rabindranath Tagore—himself both a gardener and an exponent of the language of communion with nature—writes that achieving harmony with nature is 'like gaining an instrument', not by coming to own it, but 'by producing upon it music' (quoted in Sen Gupta 2005: 66).

There is a further aspect of the idea of communion which makes it more suited to the case of growing plants than to that of hiking through the countryside. Communion is not an episodic matter, but takes place over the relatively long term; not, therefore, something that could consist in a sudden, evanescent feeling, but more a process that stretches over time—over, say, the six months from planting the squash seed to harvesting the mature vegetable. Nor is this 'process' of communion a purely temporal matter, for the continuity of communion—whether with people, other living things, or musical instruments—has, one might say, a 'narrative' structure. There is a story to be told, with a beginning and a development, when a person communes with someone or something. And certainly in the case of growing plants, such a structure is there, imposed by the needs and the good of the plants: the gardener's treatment of the plant cannot be random, but proceeds in stages, as does the development of the plant itself. When the fruit or vegetable is finally picked, a story—and so a communion—comes to an end.

This talk of structure prompts a further 'deeper' reason—beyond those encapsulated in talk of 'gift' and 'communion with nature'— why growing one's own vegetables may be significant for people. For it is not simply that this is a structured practice, it is also one which—if seriously conducted—lends a significant degree of structure and regularity to a person's life. If vegetables, or the garden more generally, are to flourish, the gardener must, as Čapek puts it, 'surrender to law and custom' (2003: 64).

Gardening is often described as a 'hobby'. But that is an unhappy term if it suggests an activity, like stamp-collecting or water-colouring, that can be 'taken up' and 'left off' almost at will—an activity, moreover, that may well be a 'bolt-on' extra to a life that otherwise goes on relatively unaffected by it. The point here, though it may be true, is not that gardening is a peculiarly time- and energy-consuming activity. Rather, it is that the specific demands and development of the 'materials'—the need, say, of certain plants to be pruned at a certain time of the year—constrain and shape the gardener's life. The life of a serious gardener is not one that, as it happens, involves some gardening. Instead, it is one partly defined by the structured, regular activities which are imposed once the decision to grow and to garden is made. When Ludwig Wittgenstein, who spent two periods as a monastery gardener in the 1920s, spoke of these jobs as answering to a 'longing for some kind of regular work', he was referring, not to regularly *paid* work, but to work that would lend a much-needed structure and pattern to his disturbed life—work, in effect, that would confer shape and unity upon his life (quoted in McGuinness 1990: 294). This special capacity of gardening to answer to a 'longing' for structure and regularity in life was noted a century earlier by Goethe in a novel whose theme is as much a garden as the relationships among the protagonists who spend most of their time in it. One of these protagonists remarks:

A tranquil eye, an unruffled consistency in doing, each season of the year, each hour of the day, precisely what needs to be done, are perhaps required of nobody more than they are of the gardener. (Goethe 1971: 224)

In effect, the life of the garden lover, not unlike that of the lover of another person, is one of 'voluntary dependence', which, Goethe tells us, is the 'best position' any of us can be in (1971: 195).

Over the last few paragraphs, I have strayed from the humble example of growing a squash. But that was anyway just an example, albeit a highly suitable one when reflecting on the inadequacy of utilitarian, instrumental explanations of the significance of gardening practices. What has emerged, I hope, is that the activities that comprise gardening are indeed significant to people for 'deeper' reasons, and possess a significance in the context of gardening that superficially similar activities engaged in outside that context could not have.

Other Garden-practices

Garden-practices, in my sense, are not confined to gardening; but nor does the expression refer, indiscriminately, to all of the countless practices that *could* be, and often have been, pursued in the garden. I once saw a dentist in India extracting teeth in the garden outside his surgery, but dentistry is not a garden-practice. Nor is torture, even if there have been real-life equivalents of 'the torture garden'— a 'symbol of the whole earth's' suffering—in Octave Mirbeau's 'decadent' novel of that name (1991: 189). The expression is not intended to be a precise one, but to indicate those practices to which, as I put it earlier, the garden is especially 'hospitable'. (I am primarily thinking, as I have throughout this chapter, of 'ordinary' gardens rather than 'great' gardens. The latter, despite or because of their scale, are not usually hospitable places. Who, wonders Sir Harold Nicolson (1968: 11)—co-creator of the Sissinghurst garden—would 'read the Sunday newspapers while sipping China tea' among the parterres of Versailles?).

Gardens are especially hospitable to practices which require, or are enhanced to the point of being transformed by, a combination of conditions: first, those such as light, open air, and sufficient space to allow for easy movement and social gathering; and second, those of relative seclusion or privacy and familiarity. It is to this combination that Sylvia Crowe refers when she speaks of the garden aiming at a balance between the 'competing ideals' of 'freedom and security' (1994: 175)—security, not only or mainly in the sense of physical safety, but in the dual sense of immunity to the outside world's intrusion and the confidence that goes with 'knowing one's way about' an intimately familiar place. This last point is worth elaborating. A garden, remarks Roger Scruton, is a 'surrounding space, not an open space' (2000: 83). He is not, of course, denying that a garden may contain areas that are open in the sense of being uncluttered, nor that, when outside my garden, I am surrounded in the sense of being encircled by trees, buildings, or whatever. Scruton's point is a phenomenological one, not a geometrical or a topographical one. What is around me in a garden with which I am familiar are 'surrounds'—an environment, that is, in which things have their place relative to one another, to the whole, and to myself. It is, one might say, an ambience of significance; and because of this, it is hospitable to practices that, outside this 'surrounding space'—in the wild, say, or in a city park—are either impossible or different in character.

Few would deny that gardens are indeed hospitable, in this sense, to many practices, including ones mentioned earlier in this chapter and on Battisti's list—from exercise and swimming to garden-parties and barbecues, from sunbathing and reverie to botanic or erotic pursuits. And the reasons why such activities are enhanced in the context of the garden might seem obvious enough: it is enjoyable to feel the breeze as one does one's Tai Chi, to watch the clouds scudding by as one lies or swims on one's back, to observe the sparrows hop on to the lunch table, to smell lavender when being kissed by one's lover—and so on. But these obvious reasons are insufficient: they do not indicate the distinctive 'tone' of garden-practices, and they merely invite further explanation. *Why*, for example, should it be good to watch the clouds above while resting? Over the next few pages, without attempting to be at all exhaustive, I try to indicate some aspects of the distinctive tone and of the 'deeper' grounds of some garden-practices.

One aspect of tone is identified by Tim Richardson when he contrasts our 'indoors' and 'outdoors' experiences of 'physical engagement' in activities. 'Indoors', typically, we go about our business 'from one carefully ordered...predictable indoor space to the next'. 'Outside', however, 'we are physically and emotionally vulnerable, and therefore more sensitive to our surroundings' (Richardson 2005: 13). His point is not that, in the garden, unlike the house, we are constantly under threat from wasps, midges, sudden squalls, and other dangers. The point, rather, is that, familiar as the environment of the garden is, there's always something going on in it, much of it novel and unpredictable. In my study, nothing ever happens—nothing to be alert to, to be 'vulnerable' to, in the sense of being open to and ready to be affected by. It's different in the garden, with its unexpected smells, sounds, and movements, and where more 'global' changes—like 'the touch of autumn' discerned one morning—bring something new to one's surrounds. It is possible, to be sure, even in one's study or dining-room, to make oneself sensitive to things—features of a painting on the wall, say—hitherto unnoticed. But in the garden such sensitivity is, as Richardson puts it, 'effortless'. Precisely because of this effortless alertness to the contingencies and novelties that affect us when in the garden, we become—as we read the papers or exercise or have lunch—'mindful', as Buddhists put it, not only of what is around us, but of our own 'physical engagement' with it, of what we are doing. This is a reason why a swim-in-the-garden is not a swim that, as it happens, is taken in a garden pool rather than the public baths. The latter is a 'predictable' environment, the former an 'unpredictable' one in which one's own body—now crossing from shadow into sunlight, now chilled by a sudden breeze, now touched by some floating leaves—itself becomes 'mindfully' salient.

A second aspect of tone, harder perhaps to articulate, belongs to activities that take place 'between', and bring together, what is 'ours' and what is not. When in my study, everything encasing me is 'mine'—walls, ceiling, floor: all of them human products that are parts of what belongs to me. When up in the hills, by contrast, nothing around, above, or below me is 'mine'—not the hills, sky, or heather: none of them human products, none of them belonging to me. Uniquely, however, when I sit or stroll in my garden, I am 'between' what is 'mine'—the lawn beneath me, the wall behind me—and what is not, the sky. As Roger Scruton, once more invoking Heidegger's terminology (see p. 58 above), expresses it, 'as we rest in the garden', we become 'aware of standing between earth and sky', and this 'enter[s] our feelings' (Scruton 2000: 83). These feelings can be profound and satisfying, ones that enhance or transform our activities. Many people, I am sure, at the end of an evening meal taken in their garden and as they look up at the stars, powerfully experience the contrast between the very human setting of the patio, with its table, food, drinks, and music, and a natural order remote from and untouched by the human. There is much talk in the literature of the garden 'mediating' between artefact and wilderness. Certainly the expression is apt for indicating the kind of experience just illustrated: for the men and women dining on their patio beneath the stars are 'between' the human and the non-human worlds, not just in the obvious geometrical respect, but—like 'go-betweens'—in virtue of bringing them into intimate relation with one another. To recall Heidegger's evocative term, the simple act of outdoor eating has 'gathered' around or into itself the different dimensions of the reality we occupy.

There are other aspects of the 'tone' of garden-practices that could be discussed, but my purpose is not to provide a complete list, only to convey that there can be no full explanation of the significance of gardens that does not invoke the 'tone' of practices to which gardens are hospitable. Nor is completeness my ambition in turning, as I now do, to some other 'deep' features of garden-practices for which the term 'tone' would be less suitable. The ambition remains the more modest one of indicating that there are indeed such features, ones which should figure in any full account of the garden's significance.

Just now, the topic was the 'mediation' by some garden-practices between the human and the non-human worlds. It is not difficult for this to modulate into the topic of practices that 'mediate' between, and forge relations between, human beings themselves. Thus, Thoreau's reflections on his famous bean-field at Walden Pond turn from the field as a 'connecting link' between places more and less cultivated to the field as a repository of past human activity, of ancient and 'unchronicled nations' (Thoreau 1886: 156 f.). As this example suggests, I am not necessarily thinking, here, of practices that are 'social', such as garden-parties, and whose serving to bring people together is obvious. I am thinking of less apparent modes of 'mediation' discernible in garden-practices, both social and solitary, including those which constitute gardening itself.

We can begin with the relations Thoreau himself experiences, between the gardener and people from the past. When we walk in a garden, even a newly created one, remarks Roy Strong, 'we walk through history' (2000*b*: 6)—and in a way we do not when walking through a newly built house. No one, five centuries ago, crossed my pristine dining-room; nor do its walls or floor contain relics of historically distant people; nor does anything grow there that people long ago planted, climbed, or sat beneath. But outside in my garden, I can be confident that where I stand, others stood centuries before, that when digging the soil, I shall reveal traces of their lives, that one or some of them planted the tree on the lawn, and that they or their descendants once climbed it or rested against its trunk. Moreover, as Strong observes, all but the most avant-garde new gardens, unlike most modern houses, 'evoke', without 're-creating', much earlier traditions of garden making. Indeed, it is common for garden designers, professional or amateur, intentionally to evoke, say, 'the medieval herber' or 'the English cottage garden'—and without the absurd results that parallel ambitions in the design of houses typically have. For the garden owner or visitor who has or is informed of those intentions, the effect is a relationship, however vicarious, with the doings and aspirations of people from previous centuries. The significance of the 'lost gardens' of Heligan, in Cornwall, for their restorers and visitors alike, clearly owes much to their inspiring a sense of 'connectedness' with the 'ordinary men and women who had once worked' and played there, to a sense—as one of the restorers puts it—of 'shared intimacy' with 'real people [who] had led real lives here'. (Too intimate, perhaps, if credence is given to the reports, by both gardeners and visitors, of encounters with ghosts from Heligan's past.) (Smit 2000: 46, 214).

An experience that awareness of past people's input into one's garden may induce is that of being its 'co-creator', one that can also be induced by reminding oneself of the input of one's contemporaries. It's not simply that friends and neighbours may have done their bit in making the garden what it is—by helping with the fencing or through gifts of shrubs and urns, say. When writing of the 'co-creation' of a garden, Tim Richardson (2005: 10) draws attention to 'the cumulative meanings accorded it by previous visitors', which, if assimilated by the garden owner, modify his or her perception of it. As with a painting that has become over-familiar, so with a garden: a friend's or a stranger's observation can so modify one's perspective that, it is tempting to say, one no longer sees it as the same thing. 'They're right', one might decide, 'it really does mirror the distant scenery, and that's the way to view it.' Or, to recall Richardson's term, the input of visitors may be more 'cumulative', as when a couple, having played host to their energetic grandchildren over the years, comes to regard the garden less as a spectacle to admire than a place of play and family gathering.

The experience of 'co-creation' is one of those which together inspire a sense of 'community'. It is hardly unknown, of course, for the owners of neighbouring gardens to be jealous rivals: but, as Čapek reminds us (2003: 63), gardening typically inspires 'mutuality with neighbours' as much as it does 'exclusivity'. For many gardeners, this is a rewarding and significant aspect of their practices. Something he most cherishes, reports Roy Strong (2000a: 212), is 'gardening friendships expressed in the exchange of gifts of seeds and cuttings, and in mutual garden visiting'. But the gardening 'community', of course, extends beyond one's neighbours. One thinks, for example, of how garden enthusiasts from Pliny through Pope to Vita Sackville-West have been inveterate correspondents, and how people today exchange views either directly or through the pages of garden magazines and websites.

It is not an objection to the 'communal' role of garden-practices to point out that, in the modern urban world, the garden, having once been 'a place for man to escape from the threats of nature', has become a 'refuge from men' (Adams 1991: 319). For it is arguable that civilized, humane relations among urban dwellers are not blocked, but rather facilitated by, these little 'refuges' to which they can 'escape'. Gaston Bachelard observes that where 'houses are no longer set in natural surroundings', so that everything is 'artificial', social life itself becomes artificial or 'mechanical and…intimate living flees' (1994: 27). Little urban gardens, so viewed, are less *cordons sanitaires* to ensure isolation from 'Them' than providers of the opportunities for intimacy, and the physical and personal distance from a 'mechanical' environment, that people require if they are to relate to one another as 'We'.

To draw attention to the garden as a refuge for the individual, while it may constitute no objection to the 'communal' role of gardens, is nevertheless entirely legitimate. Indeed, it is to draw attention to one of the abiding themes of garden literature. To enter the garden as refuge is to enter the eleventh-century Chinese author Sima Guang's 'garden of solitary delight', where all is 'under my own control, alone and uninhibited' (in Ji Cheng 1988: 124), or the place in R. S. Thomas's poem, 'The Garden', 'to remember love in, / To be lonely for a while; / To forget the voices of children'. It is to enter the 'green shade' of 'Fair Quiet…innocence…[and] delicious Solitude' in Andrew Marvell's 'Thoughts in a Garden', or the 'green oasis' in which Thomas Church's clients could erase 'memories of [their] bumper-to-bumper ride from work' (Church 1995: 6). As that last example indicates, some of the 'solitary delights' the garden affords—the harassed commuter's 'unwinding' with a Dry Martini by the pool, say—hardly call for comment. But it is clear that many authors have been concerned to emphasize the more 'cerebral' or 'internal' garden-practices to which the garden is hospitable—those of contemplation, imagination, meditation, memory. Clear, too, that for these authors, as for many other people, hospitality to such practices and 'delights' is integral to the significance that the garden has for them. It is, for example, under the heading of 'the pleasures of the imagination' that Addison grouped those 'internal' practices which made the garden one of the greatest 'delights in humane life' (in Hunt and Willis 1988: 143 ff.).

It would make matters pleasingly symmetrical to say that, just as our attention moved from the garden as 'mediating' between the human and natural worlds to its 'mediating' between human beings, so it is now moving from the latter to its 'mediating' between a human being and himself. And it is true, of course, that when meditating in a garden a person can be 'alone with himself', in more than the literal sense; true, as well, that his meditations, like those of Descartes, may have his own self as their theme. But 'green thoughts in a green shade' need not be thus directed, and it is not clear, either, that the garden is distinctively hospitable to reflections on the self: Descartes's meditations, after all, were often conducted in his famous stove. Indeed, I want to suggest that the style of meditation—'reverie', as I'll call it—to which the garden is especially hospitable is not at all the kind that focuses on one's self.

The word 'reverie' occurs in the title of Rousseau's final work, *The Reveries of a Solitary Walker*, and refers not, on the one hand, to absent-minded day-dreaming nor, on the other, to the disciplined exercises of cognition which Rousseau calls 'reflection' and which had always 'tired' and been 'painful' to him. It refers, rather, to a 'pure and disinterested contemplation' in which his mind is 'entirely free and ...[his] ideas follow their bent without resistance or constraint' (Rousseau 1992: 91 ff.). Far from being obsessively introspective, reverie enables Rousseau to 'forget [him]self'. It takes us, as Bachelard—for whom the notion is also central—puts it, into 'the space of elsewhere': it is meditation whereby 'we open the world' up to ourselves (Bachelard 1994: 184).

Reverie, I suggest, is entirely characteristic of much 'green thought in a green shade'. As often as not, the thoughts of a man or woman sitting or strolling quietly in a garden alight now on this, now on that, fluidly following their own bent, not fixated on a particular issue, object, or person, and not regimented by their owner into a chain of reasoning or towards a set purpose. Here is an example of reverie that Bachelard gives. He notices a bird's nest in his garden, and is struck by what a 'precarious thing' it is: a thought that sets him to 'day-dreaming of security', and arouses in him a feeling of 'confidence in the world'. But the nest prompts his imagination too, evoking images of a 'happy household', of his own home as a 'flourishing nest', in which his family are as close and cosy as the birds in their nest, all the more precious for also being 'precarious' (Bachelard 1994: 97 ff.). This nicely captures the fluid, passive, uninhibited character of reverie, and of its journey through thought, imagination, and memory.

But why should the garden be especially hospitable to reverie? Is it not something we can also engage in, or surrender ourselves to, in the study or the forest? Indeed we can, but that does not contradict the garden's claim to be a place of special hospitality. It owes this hospitality to the general reasons, given earlier, for the garden's hospitality to garden-practices. Probably unlike the forest, the garden is familiar: knowing what's there and one's way around it, comfortable in the surrounds it provides, one is not liable to the distractions and intrusions, even dangers, that can obstruct a reverie. At the same time, in a garden, unlike a study, something is, as I put it, always happening: something, like the chirruping from Bachelard's bird's nest, to prompt a reverie; something, like the ominously darkening sky, to push one's thoughts or imagination in a certain direction; something, like the ray of light which suddenly illuminates an old bench, to revive youthful, perhaps romantic memories that blend with one's thoughts and images. In short, the garden offers precisely the combination of conditions conducive to reverie. That is why, surely, it is in the garden as a refuge for reverie and related 'internal' practices that so many poets and essayists have found the locus of the significance it has for them.

In this chapter, I have gone beyond the 'aesthetic' aspects of the garden on which most philosophical discussions of its significance have concentrated. Without any ambition of completeness, I have tried to identify some of the many garden-practices—including gardening itself, and ranging from outdoor eating to reverie—to which the garden is hospitable, and to bring out something of the distinctive 'tone' and 'depth' of such practices. For there can be little prospect, surely, of explaining the significance of gardens that ignores what human beings *do* in them. There is another respect in which I have not aimed at completeness: for many of my remarks are, as previously indicated, prolegomena to speculations on the significance of the garden in later chapters, including the one which immediately follows.

'The Good Life'

Our 'fundamental question', concerning the significance of gardens, ramified, I urged in Chapter 1, into that of the relation of gardens to the good life. One main reason why gardens and garden-practices matter to people is because they are appreciated as conducive to the good life. By these three words, I explained, I do not mean what the lottery winner probably does when, champagne glass in hand, he announces 'Now I can really live the good life!'—a hedonistic life packed with the pleasures and fun that money can buy. (But nor do I exclude pleasure and fun from the good life.) Nor, as I also explained, do the words refer to the life of specifically moral goodness. (But nor, of course, do I exclude moral goodness from the good life.)

In Chapter 1, I suggested that one reason for the relative neglect of the garden by modern philosophy was a more general neglect of the good life and its ingredients. This suggestion might strike anyone familiar with the febrile activity going on in the branch of philosophy known as Ethics as a peculiar one. How can Ethics be accused of neglecting the good life? To explain my point, it is helpful to look at the first, or at any rate second, work of philosophy in whose title the term 'Ethics' appeared, Aristotle's *Nicomachean Ethics*. A primary concern of that work is what the Greeks called *eudaimonia*, sometimes translated as 'happiness', but better rendered as 'human flourishing or well-being'. For Aristotle, the term refers to the final purpose, or *telos*, of human beings, to the one thing desired or aimed at only for its own sake (1999: 1097b). An intimately related concern of the book is with *ethikai aretai*, sometimes translated 'moral or ethical virtues', but more literally 'excellences of character'. For Aristotle, the human good, *eudaimonia*, is, at least in large part, an 'activity of the soul in accord with virtue' (1999: 1098a). The good life, that is, is partly constituted by its manifesting excellences or virtues of character, which Aristotle describes as dispositions towards appropriate ways, not only of acting, but of judging and feeling. A later classical author goes further: *eudaimonia* 'consists in living in accordance with virtue' (Stobaeus, in Long and Sedley 1987: 394).

Modern readers of Aristotle, noting the word 'Ethics' in his title, often find themselves puzzled by some of the things that he says—and omits to say—in the book. For a start, some of his virtues do not seem to be moral or ethical ones at all: for example, friendship. We don't think of ourselves as morally obliged to have friends, nice as it may be to have them. Again, some of his virtues have little or nothing to do with how we should treat other people: they are 'self-regarding' ones, like the self-esteem of the 'great-souled' man. Or consider, in this connection, the central Epicurean virtue of *ataraxia*, the tranquillity of the person liberated from sensual needs and intellectual anxiety. Finally, there is scant discussion in Aristotle and later Greek writers of duties and rights, and of the principles or rules which should determine our actions.

What puzzles Aristotle's readers is that, in these respects, the main concerns of the *Nicomachean Ethics* are very different from those of modern works of Ethics: for in these, it is precisely an emphasis on rights, 'other-regarding' duties, and principles of moral conduct, that dominates. The puzzle disappears once it is recognized that 'ethical' refers to something different and narrower in modern discourse: no longer to what pertains, quite generally, to 'character' and its 'excellences', but to the sphere of principle, right, and obligation.

Doubtless there have been many factors at work in the history of ideas and, indeed, of political life which explain this shift in the very conception of morality, but one is particularly germane to the concerns of this chapter. Perhaps as part and parcel of the modern predilection for individual 'autonomy', there has developed in the Western tradition what Joel Kupperman calls a 'compartmentalization of human life', a 'division between two spheres of one's life'—'on-duty' and 'off-duty' (1999: 153–5). This is the division, roughly, between a person's 'public', 'civic', or 'professional' life, subject to moral rules and considerations of duty, and a person's more 'private' life, whose conduct is a matter of individual 'choice' or 'preference'. Such a division, Kupperman points out, was alien to earlier traditions, including the Greek and Buddhist ones in which, for example, the virtue of *ataraxia* or equanimity was to be cultivated in all areas of life, even the most private.

As a result of this compartmentalization and other factors responsible for a narrowed-down understanding of the ethical, the tendency is for Ethics or Moral Philosophy no longer to concern itself with what falls outside the 'on-duty' sphere of life. Its primary concern is no longer with the quality of one's 'life as a whole'—the 'point of entry for [Greek] ethical reflection', as Julia Annas puts it (1993: ch. 1)—but with what is done at 'on-duty' moments. Providing no rights are violated, no moral principles broken, then what people do in their dining-room, bedroom, or garden is nobody else's moral business: ethical considerations do not apply. The glutton, satyr, or plastic trees enthusiast may not be someone whose taste or 'life-style' we admire—but taste or life-style, not morality, is what it's a matter of.

A related result is that reflection on excellences or virtues of character—ones which may indeed inform one's life as a whole—is no longer centre stage in Moral Philosophy. What essentially matters is whether, in practice, people honour their obligations and act according to moral rules, not what their 'dispositions' are. The great moral philosophers of modernity, like Kant and David Hume, have not, of course, ignored the virtues. But for them, they are of secondary and only instrumental importance: dispositions which it is prudent to encourage if people are to come to do what is morally required of them—promote social utility, in Hume's case, or act out of a sense of duty, in Kant's.

I share with champions of recent 'Virtue Ethics' the sense that the atrophy in 'mainstream' Ethics of reflection on the good life—on the virtues constitutive of a flourishing human life 'as a whole'—is a matter for regret. By 'omitt[ing]' a whole aspect of [earlier] moral thinking', contemporary Ethics has failed to 'engage' with a host of practical, everyday concerns, and for that reason is liable to appear 'barren' to many people (Annas 1993: 455). At any rate, in the next section, I want to reflect on the virtues—the *ethikai aretai*—to whose cultivation and manifestation gardens and garden-practices are conducive. In doing so, and in thereby reflecting on the place of gardens and garden-practices in the good life, I am recalling an earlier, but still surviving, tradition of garden writing. From Pliny through Addison to J. C. Loudon (the nineteenth-century champion of the 'small garden' for ordinary people) and Roy Strong, a prominent theme has been the garden's contribution to what the first of those authors called 'a good life and a genuine one'. For the most part, my emphasis will not be upon the specifically *moral* virtues, in the narrow modern sense, to which the garden is conducive. Still, no discussion of the garden's virtues would be complete that did not respond to the moral criticisms of the garden which have been raised by, among others, some environmental ethicists. If these are well-taken, the garden is more a harbinger of moral vices than a theatre for the exercise of the virtues. Such criticisms are the topic of the last section of the chapter.

Should I not, before proceeding, tell my readers what I take the good life actually to be? If the virtues are dispositions of character that contribute to the good life, how are we to recognize *as* virtues the dispositions which garden-practices may promote without first being told what that good life consists in? Later, I shall offer some remarks on the nature of the good life, but there is no necessity to make these in advance of discussing the garden's virtues. This is because it is not unreasonable to assume a considerable consensus both on some of the constituents of the good life and on some of the virtues conducive to that life. Few people would deny that a capacity to enjoy beauty, say, or to exercise the imagination are aspects of a flourishing human life, and few would deny, therefore, that dispositions which promote such capacities are desirable.

I suspect that the impatient demand for an account of the good life ahead of identifying the virtues of the garden betrays a misunderstanding of the relationship between virtues and the good life. It betrays, in effect, the view that the virtues are merely a means to the end result of leading the good life, rather as physical exercise is a means to a well-toned physique. But this is not at all how the Greeks envisaged the relationship. *Eudaimonia* is not some end state—a state of mind, say—that, as it happens, results from virtuous behaviour; rather, the eudaimonic life is inseparable from the exercise of virtue. This is why, for the Stoics, as reported by Plutarch, 'living viciously is identical to living unhappily' (Long and Sedley 1987: 396). The good life, one might say, is partly or wholly constituted by the virtues. If so, there can be no question of first spelling out the nature of the good life and only then proceeding to identify the virtues, for no substantial account of the good life could be given that does not already invoke the virtues. Having invoked them, one may indeed reflect on how, quite, the virtues are integral to the good life—on why, in effect, we regard them as virtues. And I shall, as mentioned, be engaging in reflection of this kind: but the proper time for this is after, not before, we have identified some virtues of the garden.

Virtues of the Garden

The author of Ecclesiastes, 'The Preacher', gloomily concluded that all his toil was 'vanity', from which nothing was 'gained under the sun' (2: 11). His toil had included the making of gardens and the planting of fruit-trees (2: 5). So, not everyone perceives gardening as contributing to the good life. But the Preacher was wrong. Garden-practices really do contribute to this, and not simply because this or that person's engagement in these is a 'choice' that 'works for him'. The 'choice' of such an engagement is an appropriate one for anyone who is concerned to live well. The primary reason for this is that, as I shall put it, many garden-practices *induce* virtues. In saying this, I don't mean to deny that the garden contributes to the good life in ways that may be unconnected with the virtues—through, say, affording a sense of physical vitality when out and about in it. Notice, though, that I write 'may be unconnected': for it is not obvious where, if anywhere, the cut should be made, among ingredients of the good life, between what is and what is not a manifestation of virtue. Perhaps it is only to someone who hears the word 'virtue' as referring to moral virtue, in today's narrowed-down sense of 'moral', that it sounds odd to describe as virtuous those practices that promote a feeling of vitality.

When I say that garden-practices 'induce' various virtues, I intend the term in several of the many senses it has (or once had)—'attract', 'invite', 'bring on', 'entail'. The garden, as I argued in Chapter 4, is 'hospitable' to various practices many of which, I now argue, invite and attract certain virtues by providing especially appropriate opportunities for their exercise. By doing so, these practices 'bring on'—cultivate and entrench—these virtues: indeed, they are practices in which properly reflective engagement *entails* the exercise of these virtues.

It will help to clarify this claim by contrasting it with what I do *not* intend by speaking of the garden's contribution to virtue. I remarked earlier that there is a long tradition of associating gardens with the virtues and the good life: but here are three views found in this tradition that should be distinguished from my claim. First, there's the idea that simply being in gardens, surrounded by plants, enhances people's feeling of well-being, especially those who are ill or depressed, and thereby enables them to live more effectively (see Brook 2003: 232). Then there's the distinctly Victorian thought that gardening, rather like taking cold showers at boarding-schools, stiffens resistance to 'temptation'. Keith Thomas (1984: 234) records the optimism, in the 1860s, that growing flowers might reduce the illegitimacy rate, presumably by taking the minds of young men and women off one another. Finally, there's the characteristically Jeffersonian conviction that the qualities developed through gardening—sturdy independence and the like—would help to produce more responsible and 'virtuous citizens' than those brought up in the cities, those 'sores on the body politic', as Jefferson described them (in Adams 1991: 287).

For all I know, there is truth in the three views just mentioned, for they are *empirical* ones testable, in principle, through psychological or sociological research. Each maintains that certain practices do, as a matter of fact, have certain consequences—temperance, responsible citizenship, or whatever. My claim that garden-practices induce virtues, however, is not intended as an empirical one. Nor, I think, have all writers in the tradition to which my claim belongs intended it as such—not, for example, Addison when he spoke of gardens as cultivating 'a virtuous habit of mind' (in Hunt and Willis 1988: 146). The denial that the claim is an empirical one may sound odd: surely, it will be said, it is at most an empirical truth that growing one's own squash, hoeing, or eating on the patio 'brings on' certain virtues. Well, it is true that someone may stick some seeds into the ground, scrape soil around with a tool, or gobble down a barbecued steak in ways that leave it open, indeed doubtful, whether any virtues are thereby involved. The same is true of other activities, such as playing chess: someone can push the pieces around without manifesting those intellectual capacities—foresight, imagination, and so on—which, nevertheless, one rightly holds to be integral to the game. For when we speak of the capacities required by chess, we are thinking of serious (not necessarily solemn), understanding engagement in the game. Similarly, in holding that certain garden-practices necessarily induce virtues, one is speaking of those practices as engaged in with a proper understanding and appreciation of what they are.

The general point, here, is one made by Alasdair MacIntyre in his influential book *After Virtue* (1982), when he writes of virtues and standards being 'internal' to practices. Scientific enquiry, for example, may contribute to further, 'external' goals, such as technological innovation. But it is judged by such 'internal' criteria as respect for evidence, honesty, and the like, not by its success, on this or that occasion, in achieving some further goal. It is no objection to this point to observe that any practice *can* be engaged in with scant regard for its standards, but this would be a fake or degenerate form of engagement. So we should, for example, distinguish proper hoeing from mere, inattentive scraping of soil with a blade; it is, one imagines, the former to which John Updike refers, in his poem 'Hoeing', when he writes of 'how many souls have been formed by this simple exercise'.

Underlying this point is a still more general one, concerning the relationship between virtue and understanding. There is a tendency to think of these as independent aspects of quite separate faculties or compartments of the mind. But this is not at all how Aristotle and other older writers on the virtues saw things. Consider, for example, the central Buddhist virtue of compassion. This is not some 'inner feeling'—a bleeding-heart emotion of pity when confronted with suffering, say—for it is manifested only by someone who understands the causes, nature, and universality of suffering. Conversely, a person only fully understands these—has only fully internalized what he purports to understand—when disposed towards certain feelings and judgements and ways of actively responding to suffering. For the Buddhist, therefore, compassion and understanding—more generally, virtue and wisdom—are inseparable.

My claim, then, is that certain garden-practices necessarily induce virtues since, when properly or 'seriously' engaged in, the engagement is an understanding one, imbued with an appreciation of what is being done: and it can only be this if, at the same time, it 'invites' and 'brings on' the exercise of virtues. To identify the virtues of the garden, therefore, is to recognize how certain virtues are 'internal' to these garden-practices. How might we set about this? Well, one could draw up a list of garden-practices, including those discussed in the previous chapter, and adumbrate, as I did in connection with some of these, the explicit or implicit understanding which informs serious engagement in them—the recognition, for example, that 'newness comes into the world' through growing things (p. 73). One could then draw up another list, this time of the virtues, and proceed to work out which virtues are induced by serious engagement in which practices.

Here is a relatively straightforward example of the procedure. Consider that pleasant garden-practice of taking time off to admire the fruits of one's horticultural labour, properly to appreciate, say, the 'pictorial beauty of flower and foliage' which, like Gertrude Jekyll, one aimed to create. Such appreciation involves understanding— for instance, of the relationship of the flower and foliage to the garden as a whole (see Chapter 3)—and, crucially, it must, in something like Kant's and Schopenhauer's sense, be 'disinterested'. It is not the pictorial beauty I am admiring if my enjoyment is due to this part of the garden's utility, prize-winning potential, or capacity to make my neighbours envious. Now, like Schopenhauer, I take 'disinterested- ness' to be a virtue: for there is surely something wrong with a person who is incapable of standing back from his or her interests so as to appreciate things for what they are, rather than in terms of their contribution to those interests.

It is not my intention in what would then become an extremely long chapter to go through the lists of garden-practices and virtues, making in each case the type of connection made in the previous paragraph. Instead, I shall work through a more complex example than the one just given and, since it is not an atypical example, venture a general conclusion about the inducement of a range of virtues by many garden-practices. Let's return to the practice of growing food in one's garden, to the squash-growing gardener of Chapter 4, where I identified some 'deeper' aspects of this modest project. In explain- ing why this practice might matter to someone, I was, in effect, articulating ways in which such a person might understand it. In what follows, I recall those ways, this time connecting them with a number of virtues which may, therefore, be viewed as induced by the practice. (There is some artificiality in taking these virtues sep- arately: but I do not ignore the relations among them, ones so close, arguably, as to confirm the Socratic doctrine of 'the unity of the virtues', according to which the exercise of any one virtue requires that of the rest.)

One thing that our squash grower understood was the needs of his plant, what is required if its own good—its development into a healthy specimen of its kind—is to be realized. For this understanding to be more than mere intellectual assent to certain propositions— for it to 'penetrate', as Buddhists like to say—it must modulate into an attitude of care towards the plant. The grower is one who, at least when in his greenhouse, 'substitute[s] for the care/Of one querulous human/Hundreds of dumb needs' (R. S. Thomas, 'The Garden'). To exhibit care, whether towards querulous people or dumb life, is to manifest a virtue. It is a virtue that stands close to that of respect for life. Like the language of 'communion with nature' (p. 73), that of respect for non-sentient life has been debased by 'New Age' extravagances: but when, for instance, the Japanese priest Zōen exhorted the landscape gardener to 'maintain an attitude of reverence and respect', he was rightly urging that in our dealings with non-human living beings, it is their good and not simply our own that should guide us (Zōen 1991: §63).

A second thing our squash grower understood (p. 75) was that submission to the discipline of caring for this plant and his garden as a whole imposes a structure and pattern on a life that might otherwise be lacking in shape and unity. This 'voluntary dependence', as Goethe called it, might not itself be a virtue, but is a pre-condition for the self-mastery and self-discipline that figure prominently among the virtues enjoined in Greek and Buddhist traditions—virtues whose exercise is imperative if a person's life is not to drift or fragment. The virtue of care for life in the garden modulates, one might say, into care of self. This theme of the relationship between submission to the discipline of the garden and lending unity and structure to one's life can be heard in the following passage by Mirabel Osler, here writing not of kitchen gardening, but of viticulture:

To find a garden that is bonded to a working production ... has an irresistible attraction ... There is something whole and satisfying in seeing the physical bits of someone's life coming together ... The vineyard garden is a way of life. (Osler 2001: 165)

The grower understood, third, what Pollan intended in referring to the mature squash as a 'gift', the product not just of horticultural labour, but of 'grace' (p. 73). But properly to appreciate this, to be 'penetrated' by such understanding, is to exhibit a virtue which many writers have seen the garden as inducing—humility. In recognizing

'grace beyond the reach of art', wrote Pope, the gardener is freed
from the pride that 'conspire[s] to ... misguide the mind' (Pope 1994:
8 f.). And, back to vegetables, the 'simple wisdom' learned by Louis
XIV's head kitchen gardener at Versailles, taught him that 'everything ... should be performed as an act of humility', and not in the
manner of his royal master, with his 'insatiable will to make the world
conform to ... his dreams'. (So, at least, writes Frédéric Richaud
(2000) in a charming novel based on the life of De la Quintinie.) The
humility in question, here, is not, of course, Uriah Heepish sub-
servience or the self-effacement of a 'blushing violet', but an attitude
and stance which reflect what Iris Murdoch calls a 'selfless respect
for reality' (1997: 338) and an appreciation of the severe limits on the
capacity even of a Sun King to make that reality 'conform' to one's
purposes. (I return to humility in Chapter 8.)

Humility is closely related to hope. To engage in a project with
the understanding that its outcome is only partly in one's own
hands, but without any trust or confidence in the co-operation of
the world—in 'grace'—would be futile. Hope, therefore, is a virtue
induced by the same understanding of garden-practices as humility
is. Like humility, moreover, it is a virtue that has figured prominently in garden literature. 'What most entrances me about gardening', writes Roy Strong, is 'hope' (2000*a*: 213). Hope is included in
the three so-called theological virtues, but is it really a virtue? Not
when it consists in merely wishing for things, or in optimism that
flies in the face of the facts. But when, as in the Pauline sense of
the word, it takes the form, as John Cottingham puts it (2003: 75), not
of 'a cognitive attitude of expectation that outcomes will be ...
favourable', but 'an emotional allegiance' to 'the power of goodness'
in one's intercourse with nature and other human beings, it is what
might be called a 'foundational' virtue—a pre-condition for the
exercise of any others. For unless confidence is invested in the power
of virtuous practices to be conducive to the good life, there could
seem no point to them. Hope is not simply induced by this or that
garden-practice, like growing squash, but pervades the very ethos of
gardening. To make a garden is to engage in a planned, demanding,

long-term enterprise, one peculiarly sensitive to the slings and arrows of fortune. 'Gardeners', observes Karel Čapek (2003: 167), 'live for the future': they want to see, for example, how those birch-trees will be in several years' time. This is why the garden requires not only the investment of effort, but allegiance to the thought that 'the true, the best is ahead of us... [and] each successive year will add growth and beauty'.

So, in the above ways, growing one's own squash may indeed induce a cluster of virtues. Little turns, any more than it did in Chapter 4, on this particular example, and it is not difficult to think of other garden-practices, not necessarily those of gardening itself, which induce virtues already identified: eating outdoors, for example, when experienced in the manner described on p. 79, invites humility. There will be different virtues induced by further garden-practices: equanimity and 'disinterestedness' in the case of reverie, perhaps, or friendship and solidarity in the case of the more 'communal' practices discussed in the previous chapter.

Instead of spelling out these many possible connections, I make a general observation on the virtues mentioned in this section that are induced by garden-practices: each of them belongs to the wider economy of virtue that Iris Murdoch called 'unselfing', a process of detachment from absorption in what peculiarly concerns one's own interests and ambitions (1997: 385). (This may sound odd in the case of the virtue of self-mastery; but a central ingredient of such mastery, for the Stoics and others, was precisely mastery over one's selfish or self-centred desires.) It could indeed be argued that it is not just these virtues, but all virtues, that are marked by distancing oneself from the demands of, as Murdoch labels it, the 'fat, relentless ego'. The Buddha, for one, argued this: each virtue contributes in its own way to a transformation from the 'conditioned' state of a person in the grip of 'the conceit "I am"' to an 'unconditioned' state liberated from that conceit (see Cooper and James 2005: ch. 4). Be that as it may, it is surely under the umbrella of 'unselfing' that the virtues of 'disinterestedness', care and respect, self-mastery, and humility and hope may be sheltered.

I end this section by taking up a question heralded at the end of the previous one. While it was illegitimate, I argued, to insist on an account of the good life in advance of identifying virtues, it is perfectly reasonable to ask, once they are identified, how it is that they contribute to or are integral to the good life—to ask, in effect, why they should be counted as virtues. A popular answer to this question has been: 'Because the virtues make for happiness'—and it is certainly happiness that many garden writers have identified as the *telos* of gardening. 'The purpose of the garden', in Gertrude Jekyll's 'creed', 'is to give happiness' (1991: 91). But this answer looks to be either questionable or unhelpful. If happiness is measured, as it tends to be these days, in terms of contentment and pleasure, then it is hardly obvious— and certainly not a necessary truth—that virtuous living yields happiness, or that the vicious person is doomed to unhappiness. If, on the other hand, happiness is identified with the Greeks' *eudaimonia*, it appears tautologous to say that the virtues contribute to happiness, for the eudaimonic life, as noted earlier, was defined as life in accordance with virtue. (How different *eudaimonia* is from what is ordinarily thought of as happiness is demonstrated by the Stoic maxim that a virtuous man remains eudaimonic even on the torture rack.)

A different answer to our question invokes, not happiness—not directly, at any rate—but *truth*: life in accordance with the virtues is 'in the truth', manifesting proper recognition, that is, of the place of human existence in the scheme of things. Thus, Aristotle's ground for maintaining that the good life is one led in accordance with virtue was that only thereby is it a life 'in accord with reason', for the virtues are excellences of reason and understanding (1999: 1098a). For the Buddha, 'complete enlightenment' is inseparable from 'release' from the 'grasping' that underlies the vices (*Samyutta Nikāya* 22. 26). Such remarks testify, of course, to the intimate relationship between virtue and understanding which I emphasized earlier—a relationship, indeed, that enabled the making of connections between garden-practices and the virtues.

I return to this answer, and to the relation between living 'in the truth' and happiness, in Chapter 8, in the context of discussing the significance of the garden. Postponement of the issue, however, need not delay acceptance of the main point of this section. That we may be unsure why, quite, the virtues are virtues and how, quite, they contribute to the good life, does not mean that we are unable either to identify virtues or to appreciate that a life in accordance with them is a good life. Nor, therefore, does it mean that we are prevented from recognizing certain garden-practices as inducing virtues and as constituents, thereby, in the good life.

Some Garden Vices?

It was his impression of American gardens, during his years in the United States, which helped to confirm the theoretician of modern jihadist Islamicism, Sayyid Qutb's, negative view of the West. He saw these well-tended, well-stocked yards as symbols of a selfish individualism and materialism inimical to truly communal ideals. So, not everyone perceives the garden as an arena for the exercise of virtues. Actually, Qutb's perception was unusual, it being more common for visitors to remark on the relatively self-effacing and uniform character of suburban American gardens. But even if his perception had been more orthodox, his charge need not unduly disturb the champion of the garden. For it is not one against the garden *per se*, but against some gardeners in some social contexts. There are gardeners who strive to out-do their neighbours and flaunt their wealth, but there are many who do not.

Less speedily disposed of are criticisms, also of a moral kind, that are levelled against the very enterprise of garden making. In this section, I discuss two of the most familiar, both emanating from an influential style of environmental ethics. I call them the 'dominion' and 'deception' charges. To garden at all, it is maintained, is both to adopt an intolerable stance of human dominion over nature and to create a deceptive version of nature. The two charges tend to be levelled against different kinds of garden—'formal' or 'artificial' and 'informal' or 'natural', respectively: but sometimes they are levelled indiscriminately, against gardens as such. We shall see, at the end of this section, that the two charges rest on a common ground that reflection on garden-practice might serve to call into question.

Let's begin with the first charge. 'There is a form of environmentalism', writes Tom Leddy, which holds that 'the main problem is that man is messing up nature' and should, therefore, be kept apart from it (2000: 9). Gardens, he adds, 'will never make that kind of environmentalist happy'. For these environmentalists, not to keep apart from nature and instead regard it as a 'resource', a place to be transformed for our benefit, is a form of human chauvinism comparable, it is sometimes alleged, to male chauvinism, with nature being 'viewed as a wife or mistress', as 'objects' for scrutiny and satisfaction (see Ross 1998: 80).

A main inspiration for the dominion charge is, unsurprisingly, Thoreau. Despite his exhorting the soil to 'say "beans"', his own bean-field at Walden Pond (encountered earlier on p. 80) seemingly caused him pangs of conscience. 'What right had I to oust johnswort and the rest?', thereby treating the field as his 'property' and becoming 'degraded' in so doing (Thoreau 1886: 153, 163). Certainly, Thoreau and others who level the charge of dominion can readily find ammunition in the garden literature. 'In a garden', wrote one eighteenth-century figure, 'a man is lord of all, the sole despotic governor of every living thing' (in Thomas 1984: 238). For the Scottish captain who created the Villa Taranto garden on Lake Maggiore, two centuries later, the land he bought was enemy terrain to be conquered: 'we attacked the woods first...getting rid of...deformed chestnuts and some unattractive pines...Before long I had installed a Deccaville railway, the type we had used in Salonika in the 1914–18 War', invaluable for removing hostile trees (in Wheeler 1998: 153). And in case anyone thinks that dominion is a uniquely Western attitude, here is how Robert Fortune, the Victorian plant collector, recalls his experience of dwarf tree cultivation in China: the practice is to 'make nature subservient to... whim...a one-sided dwarf tree is of no value in the[ir] eyes....Nature generally struggles...until...exhausted, when she...yields to the power of art' (in Wheeler 1998: 173 f.).

Presumably, the charge is not the implausible, empirical one that all gardeners are possessed of an aggressive, 'despotic' mentality. It would not be difficult, after all, to counter the remarks just quoted with testaments to some gardeners' genuine concern for 'sustainability, durability, [and] integrity with the earth' (Brown 2000: 315), to a spirit of co-operation, not conquest. This point might be conceded by a critic of dominion who then confines his charge to the mentality of makers of 'formal' gardens. Schopenhauer, for example, absolves the designers of 'English' and 'Chinese' gardens, who allow 'the will of nature' to prevail, and restricts his invective to 'French' gardens—'tokens of [nature's] slavery'—in which 'only the will of the possessor is mirrored' (Schopenhauer 1969: 404 f.). But even with this concession, the charge remains uncompelling. It can hardly be seriously maintained that the psychology of everyone who creates a parterre, a rectangular pond, or intersecting *allées* is that of someone who regards himself as 'governor of every living thing'. Indeed, it was argued in the preceding section that garden-practices, when pursued with understanding, necessarily induce virtues inimical to the despotic attitude. It would be implausible to hold that these practices are never pursued with understanding, even by such exponents of the 'formal' approach as André Le Nôtre and the designers of the Alhambra gardens.

Nor should another empirical charge be taken too seriously—this time, to the effect that the control necessarily exerted over natural life in the garden is bound to spill over into a similarly controlling attitude towards 'wild' nature. Most rose pruners do not, to my knowledge, feel an itch to lop branches off the trees they walk amongst in the forest. People who make this charge are as guilty of exaggerating the similarities between experience of the garden and experience of uncultivated nature as the 'assimilationists' criticized in Chapters 2 and 3. If these kinds of experience are as distinct as I argued they are, there can be no presumption that attitudes imbibed in the garden will extend to what lies beyond its walls.

The dominion charge, then, must take the form, not of empirical claims about the psychology of gardeners, but of the accusation that, integral to the very enterprise of gardening, is an 'anthropocen-tric' stance that denies the 'intrinsic' or 'inherent' value of nature, reducing it instead to something that is there 'for us'. But there is no reason for those who indulge in talk of nature's 'intrinsic' value to suppose that gardening represents such a denial and reduction. That a gardener 'uses' living things as pot plants or for shade no more entails a denial of their intrinsic worth than his 'using' human beings as plumbers or for delivering letters entails a denial of *their* intrinsic worth. Nor need the act of gardening imply the thought that the land on which people garden is there 'for us'. There may be gardeners who hold that evolution had as its final purpose the flour-ishing of we human beings or that God fashioned nature for our benefit, so that, as St Thomas Aquinas inferred, 'it is no wrong for man to make use of [it] ... in any ... way whatever' (*Summa Contra Gentiles*, III. ii. 112). But these are few and far between, and nothing in most garden-practice is expressive of any such views. If it is never-theless insisted that any use made of nature necessarily betrays an anthropocentric stance, then the reply is that anthropocentrism is not a sin, but a pre-condition of human existence. For it is hardly gardening alone that requires making use of, and 'interfering' with, the natural world. So does just about every practical human enter-prise, including building houses and manufacturing the planes and computers that enable critics of anthropocentrism to travel to conferences and publish their criticisms. Were the garden merely a trivial, decorative adjunct to human life, then perhaps the degree of 'interference' with nature it requires would be an objection to it. But it is part of the purpose of this book to show that the garden repre-sents an altogether more 'serious' contribution to life.

I said that *most* garden-practice is uninfected by the chauvinism of which gardening is accused by dominion critics. For it is import-ant to recognize that the charge may be warranted against certain garden-practices in virtue, essentially, of what such practices repre-sent or symbolize. A good example, already alluded to in the context of the 'art-and-nature' account of garden appreciation in Chapter 3, is the kind of topiary that the authors there cited spoke of as an 'imposition of a human view onto nature'. Their animus is particularly directed against 'fantastic representational topiary'— shaping yews into, say, squirrels or sex organs. By making us 'see a natural organism as another form, a thing that it is not', by failing to 'capture the natural aspect of the tree in any appropriate sense', such topiary is in effect 'emblematic' of a cavalier attitude towards the natural world that invites such labels as 'whimsical', 'insensitive', even 'despotic' (Brook and Brady 2003: 130, 136). While it would too sweeping to proscribe all 'fantastic representational topiary'—after all, an ironic topiarist might create something grotesque precisely in order to make the point our two authors are making—the point is well-taken. The topiary being criticized is not simply an 'aesthetic affront' but, in the terminology of the previous section, instantiates a failure to exercise certain virtues—of care and humility, say—induced by garden-practices engaged in with reflective understanding.

Plate 5. Topiary structure, Botanical gardens, Peradeniya, Sri Lanka

No one is likely to be deceived into taking a topiary squirrel or phallus for the real thing; but the charge of deception, to which I now turn, is one often levelled against the garden. The criticism goes back at least to Hegel, who could not bear to look twice, he says, at an 'English' or 'Chinese' 'picturesque garden'—one which 'tries to imitate nature'—since it 'pretends to be and to mean something in itself'. Since, in gardens as in all art, 'man is the chief thing', they should proudly announce their artificiality and not dissemble as nature (Hegel 1975: 248, 699). (Here, Hegel is at odds with Kant, who endorsed the 'English taste in gardens', since these, through at least seeming to be free from 'all constraints of rules', better instantiated his maxim that art should 'have the appearance of being nature' (Kant 1952: 88, 166).) The Hegelian charge is still heard in recent times: Sylvia Crowe, for example, regarded gardens designed to 'appear subservient' to nature as displaying 'a less honest approach' than that of a 'deliberate humanizing of a wild scene' (1994: 84).

Just as eighteenth-century literature supplied ammunition for the charge of dominion, so it did for Hegel's accusation of deception. For Pope, the gardener 'gains all points, who pleasingly confounds ... and conceals the bounds' (Pope 1994: 81). He might therefore have admired Burghley House, a 1797 Guide to which boasts that 'not even the man of taste ... can ... discover the distance' between nature's and 'Capability' Brown's contributions (in Adams 1991: 174). Repton was even blunter than Pope: garden art can 'only perform its office by means of deception' (in Hunt and Willis 1988: 359).

On reflection, though, do such remarks invite Hegel's complaint? Repton, clearly, did not regard his deception as dishonourable: indeed, could it count as deception in today's familiar sense of an attempt to mislead? When Mr Rushworth, in *Mansfield Park*, resolves on commissioning Repton to 'improve' Sotherton Court, it is hardly his intention that the great man will fool its visitors into thinking that the park is a natural landscape rather than the product of garden art. It is surely morally pompous or peculiarly austere to object to all art that disguises itself, not least because informed people clearly recognize that this is what it is doing. The same is true with respect to the various 'tricks' that gardeners employ to produce certain effects. If I had just bought a property, I might be annoyed to discover that the garden is smaller than its designer had made it look—by, say, placing receding-sized paving stones towards the far end of the path that snakes away from the house. But I don't feel similarly betrayed when, having bought my entrance ticket for Versailles, I discover that the grounds are not quite so vast as they first seem, Louis XIV's engineers having exaggerated the impression of distance by making the closer ponds larger than those beyond them.

This does not mean, however, that there are not questionable forms of 'deception'. Consider, for example, what Charles Elliott (2002: 139) has called the 'counterfeit neglect' practised by eighteenth-century advocates of 'picturesque' gardens, and also by designers of many Japanese gardens in order to evoke an atmosphere of rustic simplicity and age (*wabi*). I find that my enjoyment of a Japanese garden is threatened by the possibility that the 'fallen' leaves were carefully scattered an hour ago, or the 'weathered' lanterns manufactured only a week before. The difference between this and the Versailles example, I suggest, is that, in the latter case, it is a visual experience of the ponds and parterres stretching before me that pleases, an experience which remains unaffected by the discovery that the perception of distance is illusory. In the case of the Japanese garden, however, it is a certain mood that one savours. Now while moods are not beliefs, they are susceptible to changes in belief. (The mood of elation on discerning a light at the end of a tunnel soon evaporates when it is recognized to be the light of an oncoming train.) The Japanese garden can evoke the mood intended—so I find, at least—only when the assumption remains undisturbed that the patina of simplicity, age, and relative neglect is genuine.

By focusing on Repton-like 'deception' and tricks of the gardener's trade, perhaps we are a missing a deeper point to Hegel's charge of 'pretence'. Perhaps the garden's pretence is less that of passing itself off as something it isn't than one of conveying a distorted impression of nature itself. This, certainly, is the charge made, two centuries later, by the landscape architect, Ian McHarg, when he accuses virtually all gardens of being 'representations [that] are illusory when applied to nature'. For obvious practical reasons, gardens need to be 'reassuring', 'benign', usually 'cheerful' places to be in; and this, for McHarg, means that they are 'of necessity simplifications' of nature, excluding or suppressing much of the 'richness', uncertainty, liability to change, and so on encountered in wild nature. A garden is no more faithful a representation of natural landscape than an aquarium is of the ocean (McHarg 1990: 37).

With that last remark, one might agree—not because, as McHarg maintains, gardens are unfaithful or 'illusory' representations, but rather because few gardens aim to represent nature at all. While there is much talk of the garden 'making a statement' about nature, this is true only of a very limited variety of gardens. But even if it were more common for gardens to aim at representing nature, McHarg's accusation of infidelity is misguided. Certainly, any representation— a map or painting, say—is, of necessity, a 'simplification', since it cannot reproduce every feature of what is represented. But it does not, thereby, distort or delude. McHarg, in effect, is guilty of a mimetic fallacy, for he seems to think that unless a garden itself possesses various natural properties—such as liability to sudden change or awesome scale—it cannot represent those aspects of nature. One might as well argue that a Turner canvas cannot represent the terror of a storm at sea since it is not large or liquid enough to drown in. That gardens should be 'benign' places does not entail that they are incapable of representing wild places that are far from benign. One thinks, for example, of Japanese gardens' representations of volcanoes or remote mountain ranges.

I indicated earlier that the dominion and deception charges share a common premiss—the premiss, I suspect, of much contemporary environmental ethics. This is to the effect that our practical dealings with natural places, including those we are tempted to turn into gardens, should be guided, above all, by some ideal of how we should relate to 'wild' nature. Gardeners, if such there must be, should therefore 'tread lightly' on the earth, and not exert human dominion over it; gardens, if there have to be such places, should faithfully represent how nature, uncontaminated by human intervention or anthropomorphic conceptions, really is. But this is a premiss whose familiarity should not disguise its peculiarity and lack of realism. Very little of our engagement with nature, after all, is with nature in its pristine state: it is much more with farmland, lakes which are fished, cultivated woodland, and so on. Moreover, as Michael Pollan points out, the gardener, like anyone practically engaged with the natural world, must 'learn to play the hand he's been dealt', that of a species which 'finds itself living in places... where it must substantially alter the environment in order to survive' and flourish. In the light of this, Pollan floats an interesting possibility: 'What if now, instead of to the wilderness, we were to look to the garden for the making of a new ethic?' (1996: 206 f.). What if, in other words, we were to take the relationship to the natural world that gardeners have as the starting-point for reflection on our treatment of nature—on, especially, the virtues induced in that relationship— and work outwards from there towards an appropriate stance towards wilder places? That would seem a good deal more sensible than working inwards from reflections on a relationship to genuine wildernesses in which very few human beings are involved. At the very least, we should escape the confines of a myopic concentration on wilderness that often seems to yield nothing beyond blanket condemnation of human intervention in nature, gardening included.

I return briefly to the topic of the garden's virtues and the good life in Chapter 8 after discussing, over the next two chapters, aspects of the meaning of gardens, including their representation of nature, as touched on a moment ago.

Meaning

A familiar complaint one hears these days is that, for all the modern explosion of interest in gardens, an older aspect of garden appreciation has been forgotten: that of the garden's role as a bearer of meaning. The historian, John Dixon Hunt, associates the origins of this amnesia with eighteenth-century enthusiasm for gardens that aspired to look like natural landscapes instead of providing, as Renaissance gardeners did, 'emblems' of nature. 'The major loss was an idea of the garden as representing in its own way the worlds of . . . nature', both 'wild' and agrarian (1998: 273).

This is not a complaint, I hope, to be levelled against this present book. After all, its central question is that of the significance of gardens, and already I have made several observations on dimensions of the meaning of gardens—right at the start, for example, when indicating that I would be saying little about the sociological or evolutionary significance of gardens and, more recently, when discussing the garden's alleged *mis*representation of nature. Despite this, however, I have so far said little about the notion of meaning or significance in general, nor have I tried to lend order to the remarks on meaning, either my own or ones I have cited, to be found in earlier chapters. In the present chapter, I make up for those lacunas. In between saying something, in this opening section, about the general notion of meaning and saying something, in the final section, about 'the meaning of the garden', I want to distinguish the many ways in which particular gardens or garden-practices can be said to mean something. As will soon become clear, I do not think that the term 'meaning' is ambiguous; but there are many modes of meaning, many ways in which things may mean, and it will be helpful and important to sort out some of these. Helpful, because the concept of meaning is a rather dumpy, amorphous one that, on any occasion it is deployed, invites specification of the kind of meaning involved. Important, since unless distinctions are made, some confusions and bouts of shadow-boxing which are evident in garden literature will persist.

Here are two examples of shadow-boxing. Like Hunt, Roy Strong laments the demise of the Renaissance gardener's aim to convey 'ideas and meaning' (2000*a*: 14). One might expect him, therefore, to welcome the Francis and Hester (1990) volume, *The Meaning of Gardens*: instead, he brands it as 'dotty', and clearly thinks that it fails to address the topic its title indicates. To judge from his examples of gardens that, in his view, do convey meanings, such as Ian Hamilton Finlay's 'Little Sparta' in southern Scotland, Strong equates meaning something with the intention to communicate a thought or idea. Unsurprisingly, therefore, he is not going to count the following assertion of Francis and Hester as one about meaning at all: 'Meaning resides in the power of the garden to … reconcile oppositions and transform them into inspirations' (1990: 10). For, 'dotty' or not, this is not an assertion about ideas that gardeners generally intend to communicate. But it won't follow that the authors are not addressing the topic of meaning unless, wrongly, one takes Strong's mode of meaning as the only respectable kind.

Or consider, second, Hunt's own complaint over the demise of the 'emblematic' garden and its replacement by what he calls the 'expressive' garden, one closer to Romantic taste. This could only count as a complaint about the loss of the garden's role to convey meaning if meaning is equated with something like representation. Once expression is itself regarded as a mode of meaning, as some forms of expression surely should be, then the difference between Renaissance and Romantic preoccupations is not between people who do and people who do not concern themselves with meaning, but one about the kind of meaning that gardens should pre-eminently be conveying.

I return to the modes of garden meaning in the following section. By way of preparation for this and subsequent discussions, I first make some remarks on the notion of meaning in general. (See Cooper 2003*b* for a more expansive account.) It is often said that the term 'meaning' is ambiguous. In Ogden and Richards's 1923 book, *The Meaning of Meaning*, for example, no fewer than sixteen allegedly distinct senses of the word were identified. But such claims are both misleading and defeatist. We should distinguish between a term's having several senses and its having a single sense which embraces a diverse range of phenomena. And we should not be discouraged by this diversity from looking for something that unites them.

This search is facilitated, in the case of the term 'meaning', by emphasizing the intimate connection which exists between meaning and explanation. 'If you want to understand the use of the word "meaning"', wrote Wittgenstein (1969: §560), 'look for what are called "explanations of meaning"'. Meanings, after all, are what we either understand and can therefore make stabs at explaining, or what we don't understand and then ask for explanations of. The question 'What is meaning?' is usefully treated, therefore, as the question 'What counts as an explanation of the meaning of something, be it a word, a gesture, a musical phrase, a garden feature, or whatever?'

The immediate answer to that question, of course, is that any number of different procedures might count as explaining to someone an item's meaning—ranging from pointing to something to which the item refers to describing some practice, such as a ritual, to which it belongs, or from providing an equivalent to the item, such as a synonymous word, to drawing a diagram. But this variety of procedures does not exclude there being general features which they share. I suggest that there are three features which all explanations of meaning share—ones in terms of which, therefore, the notion of meaning is to be understood.

To begin with, to explain the meaning of any item is always to relate it to what is either larger than or outside itself—to what is, in one or other of these respects, 'beyond' itself. For example, I may explain a sentence's meaning by relating it to a state of affairs in the world that it communicates, or to an attitude of the speaker that it expresses; I can explain the significance of a musical phrase by showing how it contributes to the effect of the whole piece; I might explain the significance of a garden spade by describing its use in such activities as digging—and so on. But relating an item to what is larger than or outside it cannot be sufficient to count as an explanation of meaning. I might, for instance, explain a shower of rain in terms of the wider atmospheric conditions that caused it, but this is not to say anything about the shower's meaning. So we need to characterize the kind of relation whose specification is involved when it is meaning that is being explained. The best term I can think of for this kind of relation is 'appropriateness'. One only explains the meaning of a sentence by relating it to a state of affairs or to an attitude, if it is an appropriate vehicle for communicating the former or expressing the latter. The musical phrase is meaningful within the sonata if it makes an appropriate contribution to it, and the significance of the spade is explained by relating it to digging only because it is an appropriate implement for that task.

The idea of appropriateness is, to be sure, a vague one, as it needs to be if it is to embrace the diversity of more particular relationships illustrated by my examples. But it is not therefore a useless idea. What it indicates, in effect, is that meaning is a normative notion, not a causal one (which is why explaining the shower as the effect of atmospheric conditions is not an explanation of its meaning). To explain the meaning of an item is to show how the item is suited to, or legitimate, or apt for, communicating, expressing, or standing for something, or for contributing to something, for being employed in a certain way, and so on.

So far, then, we have two entirely general features of explanations of meaning: they relate items to what is 'beyond' themselves, and they do so by indicating the appropriateness of items to what is 'beyond' them. There is a third general feature: in the final analysis, explanations of meaning refer us to what philosophers such as Wilhelm Dilthey and Wittgenstein have called the human 'life-world', human 'forms of life', or simply (human) 'Life'. Meaning, one might say, is always appropriateness to Life. Something is meaningful, wrote Dilthey, 'in so far as it signifies something that is part of Life', the 'whole' to which all meaning finally 'belongs' (Dilthey 1979: 233, 236). By 'Life', he intends, roughly, those most general practices, purposes, structures of thought and feeling, and attitudes that constitute the, or at any rate, a human 'form of life'.

It needs stressing that it is only 'in the final analysis' that explanations of meaning refer us to Life: for, of course, ordinary, everyday explanations of meaning—paraphrasing a sentence, say, or indicating a gesture's role in some ritual—do nothing as grand as invoking a whole form of life. They do not need to, but that is only because, when explaining something's meaning to someone, we typically assume and rely upon his or her sharing with us a massive, background, implicit understanding of a form of life. Once that assumption is suspended, as it has to be when, for example, we are explaining a meaning to someone from a very different culture from our own, the importance of invoking a form of life quickly becomes evident. It is, wrote Wittgenstein (1969: §241), only where there is 'agreement in... form of life' that there can be shared understanding of the meanings of words, gestures, practices, or whatever. Consider, for instance, how much a Japanese informant would have to tell an ignorant European visitor about Japanese traditions and etiquette, Buddhist beliefs, East Asian conceptions of the relationship between human beings and nature or between work and leisure, to provide even a partial explanation of the significance of the winding, uneven path that leads to the tea-house and of the need for a normal-sized person to crouch in order to enter the tea-house. The ease with which I can probably explain to my new neighbour the significance of a piece of representational topiary in my garden should not blind us to the essential role, in the success of my explanation, played by acquaintance with a whole form of life. Imagine what I would have to do if he or she had just arrived from a culture where gardening and representational art were unknown.

To sum up: meaning is what is explained in explanations of meaning. Such explanations relate—more specifically, indicate the appropriateness of—the item whose meaning is to be explained to what is 'beyond' itself. While it is normally unnecessary explicitly to invoke a whole form of life in such explanations, since we rely on an implicit grasp of this, it is nevertheless in relation to such a form that, ultimately, items have the meanings they do.

The occasional example apart, nothing in the above account is confined to the meaning of gardens and garden-practices. The account will, however, serve to inform the discussion of the several modes of garden meaning that I now want to identify.

Modes of Garden Meaning

Before cataloguing the modes of garden meaning, a few preliminary remarks are in order. First, because there are many ways in which items may be appropriate to what is 'beyond' themselves—many ways, therefore, in which meanings may be explained—there are also many ways of grouping or classifying them. There is no question, here, of just one classification being correct: rather, which modes of meaning one identifies will depend on one's classificatory purposes. For some purposes, a few very general modes suffice; for others, it is useful to break a broad mode, like 'depictive' meaning, down into several component ones (iconic, symbolic, emblematic, indexical etc.). Second, it should not be assumed that the modes catalogued are sharply distinct. We may want to distinguish depiction from allusion, say, but it will sometimes be indeterminate whether an item depicts or only alludes to such-and-such. Third, even if the modes were sharply distinct, it can sometimes be reasonably disputed which of them a given case exemplifies. When, for example, Schopenhauer (see p. 101) says that 'French gardens' 'mirror' the power of their owners and are 'tokens' of the enslavement of nature, is the point that such gardens *express* certain attitudes of their designers or simply that, irrespective of intentions, these gardens are *symptomatic* of a certain relationship between human beings and nature?

Fourth, since any item may be appropriate in several ways to what is 'beyond' itself, then it may be possessed of several modes of meaning. An arrangement of rocks may depict one thing, allude to another, express a certain attitude, be symptomatic of its times, and have personal associations for the gardener. Any of these facts could properly be cited in response to a question about the significance of the rocks, for each identifies a mode in which the arrangement is meaningful. Finally, the classification I now offer does not pretend to be complete. (Indeed, as we discover in the next section, it is intentionally incomplete in one important respect.) There may be ways in which a given garden or garden-practice can be said to have significance that are not catered for in my scheme: but I hope it covers most of what people may have in mind when they speak, in various voices, of meaning in connection with gardens.

Let me begin with two modes that can be dealt with briefly— 'mereological' (from the Greek term meaning 'part') and 'instrumental'. An item has meaning in either of these modes through making a significant *contribution* to something—in the one case, that of a part to a whole; in the other, that of a means to an end. It might get explained to a visitor that a patch of nettles which looks to him to be accidental and without any significance actually contributes to the intended medieval ambience of the place. Remove the nettles and similar features, and the garden would lose its atmosphere. Or it might get explained to him that there is meaning in the apparent madness of some strange movements being made by the gardeners: they're measuring for a new espalier, and hence doing something that is an appropriate means for achieving a certain end.

The next two modes call for less rapid treatment, for they have been the topics, though not under the labels I give them, of a massive garden literature. Both modes—'depictive' and 'allusive' meaning— could be regarded as species of a wider mode, representational meaning. Indeed, the term 'representation' is sometimes used so liberally that it becomes a virtual synonym of 'meaning' itself. But it is advisable to use the term in a more restricted way, one which is more faithful to its etymology. Roughly, an item A represents B when it is appropriate for presenting B to us—to our attention, thought, or imagination, that is, rather than 'in the flesh'. What makes the item appropriate for this task varies: a tradition, a convention, a rule, or simply someone's stipulation (as when a general, explaining his battle plan over dinner, stipulates that the salt-cellar stands for the cavalry, the pepper-pot for the artillery, and so on). Even in this more restricted sense of 'representation', it is useful, at least in the context of the arts, to distinguish between what I am calling the 'depictive' and the 'allusive'.

So important in the history of the arts, especially the visual, has been the endeavour to represent things that are available to ordinary sense perception—physical objects, bodily actions, events, and so on—that the representation of such things by artworks deserves a label of its own: 'depictive' meaning. A garden, garden-feature, or garden-practice depicts something, then, when what it represents is available to sense perception—or, at any rate, something that would be thus available if it actually existed. (A garden statue can depict a unicorn or Zeus even though no such being exists.) 'Depict', here, needs to be understood in a fairly broad way—so as to allow for sounds, say, and not just visual items, to depict, and to allow, as well, for actions, not just static items like statues, to depict.

It is depiction on which popular histories of the garden focus predominantly when discussing and illustrating the garden's capacity for meaning and representation. It is under this heading that such standardly cited instances of representation as the following fall: the rocks in Japanese Buddhist gardens that represent the Isles of the Immortals, Mount Sumeru, cranes, tortoises, and turtles; the Mughal *chahar bagh* gardens, whose four squares represent the regions of the Garden of Eden divided by the four rivers of life; Stourhead's representation, through its inscriptions and other artefacts, of people and episodes from the *Aeneid*. Readers may add to the list at will—the mound in Sir Frank Crisp's garden at Henley, for instance, which depicted the Matterhorn, or the garden at a recent Chelsea Flower Show that depicted the University Boat Race. If such 'static' items can depict, then so, in my generous sense of the term, can activities undertaken in the garden. (If that sounds odd, think of mime.) *Cognoscenti* wandering around Stourhead may be depicting Aeneas's journey, and children climbing Sir Frank's mound might have been depicting an ascent of the Matterhorn.

Not everything that garden-features represent or stand for, however, is an object of depiction. Rocks in Japanese gardens represent islands and mountains, but they may also or instead, one reads, 'symbolize time in its longest imaginable extent, and therefore the impermanence of all things' (Harte 1999: 24). The universal impermanence postulated by Buddhists is not available to sense perception, and cannot therefore, in my sense of the term, be depicted. Or consider Mara Miller's sensitive account of the sunken garden at Cliveden, with its flat stones commemorating the Canadian dead of the 1914–18 War. Through the counterpoint of the stones to the rich surrounding vegetation, the echoes of medieval tradition, and other devices, the garden inspires an 'awareness of honour' and an 'awareness of fate' (Miller 1993: 144 f.). It represents honour and fate, one might say—things no more depictable than universal impermanence. (Miller's example, incidentally, illustrates how it is sometimes necessary, in explanation of meaning, explicitly to invoke general aspects of a form of life. For the force, the appropriateness, of the Cliveden garden, on her interpretation, essentially owes to its location in a whole culture shaped by the practice of war and attitudes towards death, honour, and the past.)

The examples just given illustrate what I am calling 'allusive' meaning: the rocks or the Cliveden garden have meaning, in this mode, by 'alluding' to impermanence or honour and fate. Allusion is not intended as a sharp notion, and 'allude' is something of an umbrella term, replaceable in context by verbs like 'evoke' and 'intimate'. A garden or a garden-feature alludes to 'objects', like honour, when it represents these in virtue of its appropriateness as an item to lead to or inspire thoughts or a sense of these 'objects'. Allusion is, or may be, altogether more elastic than depiction, for what is alluded to need not be dictated by convention or the intentions of the garden designer. Miller's interpretation of the Cliveden garden is just that, an interpretation—one that plausibly identifies something that the garden is appropriate to communicating, but not one that necessarily excludes other, perhaps equally persuasive, interpretations. In this respect, interpretation of 'allusive' meaning is akin to that of metaphorical meaning, for it is typical, or perhaps definitive, of 'live' or 'fresh' metaphors that there is flexibility in their interpretation. Indeed, one could say that the commemorative garden is a metaphor for honour or fate. Gardening is 'so full of metaphors', remarked the eponymous heroine of *Mrs Miniver*, 'one hardly knows where to begin' (quoted in Howard 1991: 218). And certainly it is in the area of 'allusive' meaning that talk of the garden as a repository of metaphor is most at home.

A further reason why allusion is not a sharp notion is that it is hardly determinate which 'objects' are, as I have put it, available to sense perception and, therefore, where the exact boundary between depiction and allusion lies. Give me a pencil, or some gardening tools, ask me to depict fate, and I wouldn't know where to begin. But what of Iris Origo's garden which a friend, progressing from the cosy house, through the open spaces, to the shade of an old chapel, responded to as an 'allegory' of 'life's journey' from the womb to the grave (in Wheeler 1998: 343)? Is life's journey something that we can *see* a traveller embarked upon and therefore depict, or can it only be alluded to, perhaps by depicting particular episodes of it? My own inclination is to speak with Origo's friend, not of depiction but of allegory, and to add that allegory is a species of allusion. But some readers might be otherwise inclined.

Among the 'objects' to which gardens may allude are emotions, feelings, moods, and attitudes—so-called affective states. If a rock depicting a tortoise can allude to longevity, then a statue depicting Cupid can allude to erotic desire. But there is an importantly different, if not sharply distinct, mode in which gardens may be appropriate to the emotions, attitudes, and so on which figure in explanations of their meanings. They may *express* them. For example, a design in a knot garden that represents the interwoven initials of the names of its two owners may be expressive of their love for one another. So a further mode of garden meaning will be that of 'expressive' meaning.

Aestheticians have spilled a lot of ink in the attempt to state what it is and what it is not for an artwork to express an emotion or other affective state. It is fairly clear, for a start, that for a work to express a given emotion, it is not required, as Tolstoy thought, to be the product of an upsurge of that emotion in its creator. Nor is it necessary, as Tolstoy also thought, that the work must cause that emotion to well up in those who encounter it. I can judge a garden to be expressive of melancholy or, as it may be, *joie de virre*, while knowing that its designer is a cold fish and without myself becoming depressed or jolly. It is rather harder to say, however, what expression *is*. But a clue, perhaps, is the following: where an item expresses a certain affective state, it may properly be described by the term for that state (or an adjective cognate to it). If, for example, a garden expresses a mood of melancholy, then the garden itself is a melancholy one. If raking the gravel in a Zen garden is expressive of humility, then the raking is itself a humble action.

With this as our clue, we may then say: a garden is expressive of an emotion, etc. if it is appropriate to it in virtue of the emotional term being applicable to the garden itself. What makes the emotional term applicable is, first, that the work *would be* one appropriately produced by someone *if* he were actually experiencing that emotion, and second, that it *would be* one appropriately responded to by someone *if* he were actually being affected by it with that emotion. What matters, then, is not whether garden makers and visitors actually experience a certain emotion, mood, or attitude, but whether, if they do, it is appropriate for them to do so. I don't know whether the designer of the Cliveden commemorative garden felt sympathy for the fallen of World War I and anger at the huge loss of life, nor do I know how many visitors to the garden come away with such feelings. But the relevant question to ask, when wondering if the garden expresses these feelings, is whether the garden would be an appropriate product of, and inspiration for, such feelings, if they were actually experienced. Could a garden, though, be expressive of an emotion even though no one actually experiences that emotion in its presence? Unlikely, perhaps; but one might imagine a great garden writer of the future persuading his or her readers that a crucial aspect of a certain garden has always been missed, one which, when properly appreciated, will be seen to be suited to the expression of the emotion in question.

The penultimate mode I draw attention to can be labelled 'symptomatic' meaning. Those clouds mean rain, her spots mean she has measles: the clouds or spots, that is, are a symptom or sign of something, and in virtue of a contingent, typically causal connection they have to it. (H. P. Grice (1989: ch. 14) calls this 'natural' meaning—a misleading term, perhaps, given that, as he explains, the sign or symptom can be an artefact. The anti-measles medication in her bathroom cabinet can also mean she has measles, and so be a symptom, in my broad sense, of the illness.) It is with the mode of 'symptomatic' meaning that historians and sociologists of gardens are preoccupied, and I have already given examples of claims they make: the eighteenth-century 'English garden' as symptomatic of a patriotic, anti-Continental love of liberty; the combination of 'prospect' and 'refuge' in many gardens as a sign of our distant ancestors' need to occupy terrains that enabled both vista and shelter. Examples can be multiplied at will: William Howard Adams, for instance, understands Mughal gardens as signs, *inter alia*, of 'stark political power' and of a culture in which both a desire to control nature and a confidence in 'divine connections and support' prevailed (Adams 1991: 75 ff.).

I make three remarks on the 'symptomatic' meaning of gardens. The first is that, just as the symptoms of an illness may impact upon it (as when panic attacks make the underlying condition even worse), so gardens that are an 'epitome of a culture as a whole' (Ross 1998: 107) may in turn reinforce or reshape aspects of that culture. Keith Thomas gives the apposite example of the nineteenth-century English working man's preoccupation with his allotment, both a sign of, but also a factor in prolonging, 'the relative lack of radical and political impulses among the British proletariat' (Thomas 1984: 240). Second, a word on how the notion of 'symptomatic' meaning can be accommodated by my account of meaning as appropriateness—for it sounds odd, after all, to speak of the clouds, say, as being appropriate (or inappropriate) to the imminent storm. The first observation to make is that, in the case of gardens and many other human products, these are typically suited, in more or less obvious ways, to what they are signs of. It is no accident, for instance, that the English parks allegedly symptomatic of a spirit of national independence were planted with such natives as the oak and the hornbeam, not with palm-trees or all-too-French cypresses. The second, more important point is that, strictly speaking, what is appropriate is not the symptom itself (the clouds, the measles spots, the allotment, the oak planting), but our use of the symptom in making inferences from it to the storm, illness, culture, political atmosphere, or whatever of which it is symptomatic. Because of the contingent, causal connections between symptoms and what they are signs of, the weather forecaster, doctor, or garden historian may appropriately 'read' the symptoms as indicative of the latter.

My final remark on 'symptomatic' meaning is simply to stress its distinctness from other modes I have identified. It is important to stress this, not least because of a familiar temptation, which often betrays a somewhat cynical attitude, to 'reduce' other modes to this one. The temptation is especially evident in popular accounts of human behaviour—exchanging gifts, for example, or handshakes—which, having identified the social or biological functions of such behaviour, then conclude that the behaviour is entirely to be understood in such functional terms. When we give each other Christmas presents or shake hands, we are 'really' and 'only' serving those functions, and not, as we naïvely imagine, acting out of spontaneous affection or pleasure at seeing someone. Such conclusions are illegitimate: in my terminology, they are due to a confusion between the 'symptomatic' meaning of an action (what it shows about our social structure, say) and its 'expressive' meaning (what feeling or attitude it appropriately communicates). There is no reason at all why an action cannot mean in both ways. It is true, of course, that 'symptomatic' meaning can, as it were, be exploited. Knowing that a long face is a natural symptom of depression, I may—whether

I am actually depressed or not—pull a long face in order, quite intentionally, to convey that I am depressed. Symptom has become commandeered for the purpose of expression.

These points apply in the case of garden meaning. Roy Strong—no admirer, it seems, of the former Prime Minister—suggests that 'Thatcher's Britain...one long celebration of...private property rights', found its 'perfect garden expression' in the 1980s 'explosion' of boxed parterres, laburnum tunnels, and geometric potagers (2000a: 197). But this is unduly jaded even for a non-celebrant of private property rights, for it implies that the gardens were not only symptomatic of a wide extension of these rights, but were celebratory expressions of enthusiasm for this new order. There is, though, no reason to 'read' such an attitude into the mentality of all or many of those people who enjoyed their parterres, tunnels, and potagers, reminiscent though these may have been of 'a pre-socialist Britain'. To be sure, garden owners may exploit symptoms, just as some people exploit medical symptoms. Those who have learnt that 'decking' is a symptom of affluence and chic, or that a clipped lawn is a sign of bourgeois respectability, may 'deck' their patio and shave their lawn in order to indicate how well off, fashionable, or respectable they are. But it would be unfair to accuse all or most gardeners who 'deck' and mow of this exploitative strategy.

If the 'symptomatic' meanings of gardens are typically social and cultural, then the final mode of garden meaning I mention—'associative' meaning—more typically engages with the personal and individual. Indeed, it has been suggested that the decline of the 'emblematic' garden in the early eighteenth century was partly the result of a new interest, inspired by John Locke's 'associationist' psychology, in the 'private as opposed to the public workings' of the mind (Hunt and Willis 1988: 34). We often speak, not of an item's meaning something *tout court*, but of its meaning something *to* somebody—of its mattering to a person, being important to him or her. Consider an inconspicuous garden found by most of its visitors to be neither beautiful nor interesting, but which means a lot to a certain woman. It is not, as far as she is concerned, that the garden depicts, expresses, or alludes to anything. Instead, it is, perhaps, the place where she was proposed to, or where she heard the distant sound of a piano that decided her on a career in music, or ... There could be any number of associations that the garden has for her in virtue of which it has meaning to her. It may even be that the associations cannot be properly articulated by her: that the garden, as Bachelard puts it (1994: 13), provides an 'orientation' towards a 'secret' that she is unable 'objectively' to state and communicate.

While it is typically 'personal', meaning in the 'associative' mode need not be strictly 'private' and unshared. In the Maltese writer Joe Friggieri's story, 'Two Photos', an elderly and infirm couple have their photo taken in a public garden because this is where, in happier, healthier times long ago, they were once photographed together (Friggieri 2004: 58 ff.). Or consider the shared associations that a garden, made in the style of their homeland's tradition, might have for a group of expatriates. The carrying of such associations is not, of course, a prerogative of the garden: houses, books, and beaches have them as well. It is hardly surprising, however, that authors often choose the garden—as in the Friggieri story, or in stories by D. H. Lawrence and Virginia Woolf—as the scene which prompts the memories, nostalgia, and regrets of their protagonists. After all, as argued in Chapter 4, the garden is peculiarly 'hospitable' to the exercise of reverie, memory, and imagination.

This completes my catalogue of modes of garden meaning. Some readers may be able to think of instances of meaning that cannot be fitted under any of my labels—'mereological', 'instrumental', 'depictive', 'allusive', 'expressive', 'symptomatic', 'associative'—but I hope that, with one notable exception, these omissions are not important. It is to that exception that I turn in the following section.

'The Meaning of the Garden'

Attention to the various modes of garden meaning just catalogued is important in garden appreciation. For some connoisseurs, great pleasure is to be had from identifying what garden-features depict; for other people, it may be the expressive power of certain gardens that attracts them. But we do not ask only about the meanings of particular gardens: we also ask about the meaning of the garden, where the last two words refer not to this or that garden, but have the same role as the words 'the clarinet' when a child asks what the clarinet sounds like. When used in this way, I'll write the two words thus, 'The Garden'. Indeed, the 'fundamental question' with which the book began, about the significance of gardens, can be rephrased as the question 'What is the meaning of The Garden?' And we have already met with claims, often hyperbolic and dramatic, which could be construed as answers to this question. For Gertrude Jekyll, The Garden 'teaches' or means 'trust' that 'God will give the increase' (1991: 25); for one theologian, the meaning of The Garden is also religious—'the presence of ultimacy', with which gardens are 'saturated' (Matthews 1990: 46); for one philosopher, The Garden is a metaphor for, and so has as its meaning, 'the ideal human life', one in which the 'opposites which constrain our existence' are 'reconciled' (Miller 1993: 25 f.).

A question naturally arises concerning the relationship between the meaning of The Garden and the modes of garden meaning identified in the previous section. The answer looks to be that there is not much of a connection. Attention, say, to the 'depictive', 'expressive', and 'symptomatic' aspects of garden-features, in my senses of those terms, seems remote from the kind of attention to The Garden that inspires claims like the ones just cited. After all, attention to the former aspects is characteristic of *cognoscenti*—historians, aestheticians, and so on—focused upon particular features of particular gardens rather than upon The Garden as such. And it would surely be hopeless to search for any one 'object' that is depicted by, expressed by, alluded to, signalled by, or associated with The Garden as such and which might then qualify for being its meaning.

The lack of connection between the catalogued modes of garden meaning and the meaning of The Garden will be sufficient reason, in some people's minds, to pronounce the question of The Garden's

meaning as improper. There is nothing 'beyond' to which The Garden
as such is appropriate, hence no meaning for it to bear. (Compare a
familiar attitude towards the question of the meaning of life: if this
isn't reducible to plain talk about the meanings of particular human
activities and projects, one often hears, then it's a silly one.) If, after
all, the catalogue of modes of meaning is exhaustive, how can we
even understand talk of the meaning of The Garden unless this is
reducible to one or more of those modes? Well, the question now is
whether that catalogue was complete, whether there is not a dimension of meaning that got left out, but which may now be invoked in
order to illuminate the notion of The Garden's meaning.

Before addressing this question, however, some remarks on the
expression 'The Garden' are in order. It does not refer, we know, to
any particular garden, but nor does it refer to all gardens. (Compare
'The Clarinet': someone who likes the sound of The Clarinet does
not thereby like every noise that comes from every clarinet, irrespective of who is blowing it, Benny Goodman or a monkey.) Philosophers
of biology, confronted with analogous terms in their field of interest,
like 'The Tiger', often propose the following: a statement like 'The
Tiger is a carnivore' is not about a particular tiger, or about all
tigers, since it remains true despite the existence of toothless tigers
reduced to eating plants and of soya-fed ones in zoos. Rather, it is
a statement about 'normal' tigers, ones which are 'typical' of their
natural kind or species. This proposal won't do, as it stands, for The
Garden (or The Clarinet), since gardens (or clarinets) do not constitute a natural kind or species. But I make the following proposal,
which is in the spirit of the one made by philosophers of biology: a
statement of the form 'The meaning of The Garden is...' is not
about a particular garden or about all gardens, but about 'exemplary'
gardens. Jekyll's claim, for example, could now be construed as
telling us that the meaning of exemplary gardens is trust in God's
gift of increase. And this claim could be true even if there are
gardens—ones which are not exemplary and do not approximate
sufficiently to those that are—which fail to inspire or convey such
trust. (Think of a garden so polluted that nothing grows in it.)

So the question now becomes one about exemplary gardens—those which stand to The Garden in a manner roughly analogous to the relation between 'normal' or 'typical' tigers and The Tiger. Is there a mode of meaning, so far missing from our catalogue, that all exemplary gardens have, irrespective of the differences among their 'depictive', 'allusive', or whatever meanings? The answer, as we can learn from Nelson Goodman, is 'Yes'. In Goodman's writings on 'the ways of meaning' of artworks, especially ones that do not purport to represent anything, special attention is paid to what he calls 'exemplification'. Items exemplify properties, he says, when 'they both possess and refer to' these properties (Goodman and Elgin 1988: 19), though not in virtue, as I defined the terms, of depicting or alluding to these properties. Thus a tailor's swatch may exemplify the colour and the kind of a material, and the Villa Aldobrandini, in Frascati, may exemplify the Roman Baroque style of garden design. Possession of and reference to its properties is a distinct way in which something is appropriate to properties, hence a distinct mode of meaning.

Three points should be made about exemplification, in order to avoid possible misconceptions. First, an item does not exemplify every property or feature it possesses. The swatch is, say, two inches square and red, but it does not exemplify its size as it does its colour; the Villa Aldobrandini was completed in 1603, but it does not exemplify the property of being completed in that year. To exemplify something, the item must 'refer to' it: it must, to borrow Kant's phrase, 'body forth' what it exemplifies, so that the latter, in Goodman's words, 'comes forth'. Second, it is perfectly possible for a work both to represent or depict and to exemplify—indeed, perfectly possible for it to depict and exemplify the very same properties. The grounds of Stowe may exemplify some of the virtues, such as harmony, that their Temple of Ancient Virtue and Elysian Fields actually represent. A garden may exemplify fecundity and also allude to it, perhaps through topiary that depicts horns of plenty. But, as these examples show, the two ways of meaning the same properties need to be distinguished. Finally, exemplification should be distinguished from expression, as I employed the term. True, in order for a garden to express an emotion, like melancholy or joy, it must be melancholic or joyous. But it is only metaphorically so: being insensate, the garden is not literally melancholic or joyous. 'Exemplification', as I use the term, applies only to cases where the item literally possesses the features it exemplifies. (Here my terminology differs from that of Goodman, who speaks of expression as 'metaphorical exemplification'.)

While exemplification is not the monopoly of artworks that are non-representational (or whose representational role is secondary), or indeed of art more generally, it is an especially important mode of meaning to consider in connection with such works. Goodman himself has devoted much attention to the exemplifying meanings of music, abstract painting, and architecture, and many of his remarks, especially those on architecture, translate easily into the domain of horticulture (Goodman and Elgin 1988: ch. 2). Thus it is surely essential to an understanding of Thomas Church's gardens— swimming-pools, barbecue patios, parking bays, and all—to recognize that, as I put it earlier (p. 63), they 'instantiate', that is exemplify, the practical functions that he deemed it the purpose of the twentieth-century garden overtly to perform. Again, it may be as important to the appreciation of a Modernist garden as it is to that of some Modernist buildings to recognize how it exemplifies—draws attention to, 'bodies forth'—its own structural features.

A last remark before returning to the meaning of The Garden is that exemplification comes in degrees. Just as some tigers (fit, four-legged, meat-eating ones) are better exemplars of their kind than others (mangy, three-legged, toothless ones), so, of two gardens that exemplify a certain style, one may be more exemplary than the other. The Villa Aldobrandini, for example, is generally taken to be a truer exemplar of Roman Baroque than some later ones, such as the Villa Falconieri. Properties which, although they are exemplified by both gardens, may 'come forth', be 'bodied forth', more saliently or vividly by the first villa than by the second. All exemplary gardens, one might say, are appropriate to what The Garden means, but some are more appropriate than others.

Plate 6. Geoffrey Bawa, 'Modernist' garden and house, Sri Lanka

With the 'missing' mode of garden meaning now joining the catalogue, I propose that the meaning of The Garden—of exemplary gardens—is something that The Garden exemplifies or, to introduce some terms I invoke in the next chapter, that it embodies and serves as an epiphany of. When, for example, Jekyll proclaims that meaning of The Garden is that 'God will give the increase', this is to be construed as saying that The Garden exemplifies, embodies, or is an epiphany of, God-given increase. When it is held that The Garden's meaning is an 'ideal human life' in which opposites are 'reconciled', that is to be taken as maintaining that The Garden exemplifies a reconciliation of opposites (nature and culture, say, or change and continuity). So construed, claims about the meaning of The Garden tell us that it has certain properties which, in exemplary ways, it 'bodies forth'. Both of the claims just mentioned may, of course, be contested: but they conveniently illustrate, not only the form that, on my proposal, claims about the meaning of The

Garden have, but also the *kind* of property that The Garden must exemplify if there is to be such a thing as its meaning. At any rate, it has got to be a kind of property—unlike that of being Roman Baroque, or of having certain practical functions—which is not confined to just some exemplary gardens, but one which they all have. I devote the next chapter to exploring what this property might be and to articulating what I think it is.

Symbol and Epiphany

The meaning of The Garden, if there is one, is what exemplary gardens exemplify. The Garden is appropriate to what it means through exemplifying it. Such was the thought announced at the end of the previous chapter and the one I develop in the present chapter. First, however, some further remarks on exemplification and related notions. Nelson Goodman, from whom I have borrowed the terminology, tends to provide fairly sober examples of this mode of meaning: the swatch that exemplifies its own colour, the painting that exemplifies a certain style, and so on. It is not his intention, however, to exclude cases of exemplification closer in character to the headier claims about The Garden's meaning that I cited in Chapter 6. He gives, for instance, the putative example of a 'chain of reference' through which a Modernist church, which depicts sailboats that exemplify 'freedom from earth', which in turn exemplifies 'spirituality', thereby exemplifies this spirituality (Goodman and Elgin 1988: 42). It is not inappropriate, therefore, to relate Goodman's notion to an older one—the Romantics' conception of symbols.

Writers such as Goethe and August Schlegel were at pains to distinguish symbols from other vehicles of meaning, such as signs, allegories, and symptoms. A symbol is related to what it symbolizes neither in the merely contingent, causal manner of a symptom to the condition it indicates, nor in the purely conventional manner of signs, such as words, to what they stand for. Instead, symbol and symbolized are somehow 'fused', or as Wilhelm von Humboldt put it, in 'constant mutual exchange' with one another (in Molesworth 1995: 413). Such remarks are reasonably interpreted to convey that symbols have meaning through exemplifying what they mean: an item is 'fused' with the properties it refers to in virtue of possessing those very properties. But the remarks convey something further—what one commentator has called the 'transcendent pretensions', or 'striving toward transcendence [and] gesturing to the supersensible', which Romantic writers attributed to symbols (Allison 2001: 256, 260). Actually, that comment was made with respect to Kant's view of symbols—'aesthetic ideas', as he called them—as striving to 'body forth' what 'language ... can never get quite on level terms with or render completely intelligible' (Kant 1952: 176). Symbols are in 'mutual exchange' with what they symbolize, since they enable us to understand, 'get a handle on', or become attuned to, the latter. The symbol is not an aesthetically pleasing, but otherwise redundant device for drawing attention to what we are already and independently able fully to grasp and articulate. Rather, it is indispensable to enabling a sense for, an attunement to, what it symbolizes.

The mode of meaning of an item credited with the 'transcendent pretensions' of the Romantics' symbol deserves, perhaps, a more resonant label than 'exemplification'—'embodiment', for instance, or, better still, 'epiphany'. For if Goodman's discussion recalls an older, Romantic one, this in turn invoked a still older tradition, according to which something 'spiritual', and possibly ineffable, 'shows itself' in sensible form, and thereby enables human beings to gain a sense and understanding of it. In this tradition, the ordinary—the sensible, the worldly—is to be experienced as an epiphany. (To take the most obvious example: God 'epiphanizes' in human or other worldly forms, or perhaps as the world as a whole.) From now on, then, I use whichever of the terms 'exemplification', 'embodiment', and 'epiphany' is best suited to the particular context of discussion.

Our question, then, is this: of what, if anything, is The Garden an epiphany? In the next section, having listened to the testimony of one great artist and garden lover, I consider a relatively 'modest proposal' by way of an answer to our question. The proposal, I will suggest, is fine as far as it goes, but this isn't far enough. What is modestly proposed as the meaning of The Garden—as what it exemplifies—turns out, rather in the manner of Goodman's church–sailboat–freedom–spirituality chain, to exemplify something further. It is this latter, I argue in the final section of the chapter, of which The Garden should be seen as an epiphany.

Before turning to these matters, however, I first want to consider an issue of a broadly methodological kind which, unfortunately, is too often ignored by those, like myself, who want to make large claims about the meaning of The Garden. For two reasons, the meaning of The Garden must be 'available' to people for whom exemplary gardens matter, something of which they have at least an implicit recognition. This must be so, first, if the appeal to what The Garden exemplifies or embodies is to explain or illuminate why The Garden matters to people. Were The Garden an epiphany that no one recognizes as such, then nothing would get explained as to the significance it has for people. Second, it belongs to the very notion of exemplification that, from an exemplifying item, something 'comes forth' or is 'bodied forth'. But these are relational expressions: what comes or is bodied forth must do so to or for people. If the item fails to refer people to anything, to render anything for them, then it fails to exemplify.

So the question of the meaning of The Garden raises the issue of how one tells that this meaning is, as it must be, recognized by people. What reasons could there be to maintain that people for whom The Garden matters *take* The Garden to embody what I, or some other writer, propose? Somebody might respond: 'That's easy! Ask people!' Unfortunately, it cannot be that easy: as we saw in Chapter 4, when discussing the 'deep' significance of particular garden-practices, like growing one's own food, a phenomenology of experience is not to be judged by its correspondence to what people actually say about their experience. The phenomenological task is an interpretative one, not that of reporting what people are anyway perfectly able to tell us for themselves. This point is especially germane when, as now, we are addressing a kind of meaning with which, to recall Kant's statement, language may be unable to 'get on level terms'. A more plausible response is: 'Tell people what *you* think The Garden embodies, and see if they agree!' Certainly, it would count against an interpretation of The Garden's meaning if it didn't 'ring true' for people, if there were no acknowledgement by them that, yes, this helps to articulate something they've long felt or dimly discerned. But it's important to realize that it will be the whole interpretation, contextualized, explained, and defended— something that might take up a whole book—not some staccato statement of it, to which people's acknowledgement or lack thereof would be relevant. Go up to your next-door neighbour and say 'I think The Garden is an epiphany of an ideal of life in which opposites are reconciled. How about you?', and you are liable to draw a blank.

Someone who proposes that The Garden is an epiphany of some-thing—call it X—need not, however, be without strategies to lend support to the proposal. One such strategy is to focus on a number of, so to speak, hyper-exemplary gardens—ones which are found, by many of those who know them, to epitomize The Garden in their various ways. For if these gardens seem to fit the proposal especially well, this might help both to explain people's response to them in terms of the proposal and to provide some grounds for extending the proposal to all exemplary gardens. All of these embody X, if less vividly and unmistakably than the hyper-exemplary ones do. A very different strategy is to appeal to, and reflect upon, the testimony of those—poets and other writers—who are in the business, as it were, of trying to articulate what The Garden might embody, especially when it is testimony that has resonated with many readers. Pope's famous lines, in his Epistle to the Earl of Burlington, on 'the genius of the place', for example, surely evoke a conception of The Garden as an epiphany. For Pope, 'the genius of the place' does not refer, as it does for many later writers, to the ambience or natural setting of a garden: rather, it is that which 'Now breaks, or now directs, the intending lines' and 'Paints as you plant, and, as you work, designs' (Pope 1994: 81 f.). Palpable, here, is a sense of The Garden as both a response to and an exemplification of something beyond the control and invention of human beings. And so too, to turn to a very differ-ent literary work, in Frances Hodgson Burnett's children's novel

The Secret Garden. The 'magic' that young Colin senses to be at work in this garden is equated, by the woman who then speaks to him, with 'Th' Big Good Thing' that 'goes on makin' worlds' (Burnett 1994: 277).

A related strategy is to recognize the import of traditional discourses that have been peculiarly prominent: for it may be that these, without directly registering a sense of The Garden as an epiphany of X, could nevertheless be understood as gesturing towards this. Now nobody, surely, could be unimpressed by the scale of *theological* discourse about The Garden. Dorothy Frances Gurney's contention that 'One is nearer God's Heart in a garden / Than anywhere else on earth', lampooned though these lines often are, belongs to a long tradition of perceived intimacy between The Garden and God. So, more pithily, does Alan Titchmarsh's remark, apropos his 'garden make-over' team in a popular television series of the same name, 'Landscaping from *Ground Force*, landscape from heavenly force'. Sometimes in this tradition, it is made quite explicit that The Garden does not simply require God's assistance, and is not simply a place in which the mind turns to God, but is a divine manifestation or epiphany. Here is how the great thirteenth-century Sūfi poet, Rūmi, expresses it:

> In the garden are hundreds of charming beloveds
> And roses and tulips dancing around
> And limpid water running in the brook
> And this is a pretext—it is He alone.

> (in Schimmel 1992: 70)

Now such a specifically theistic account of the significance that The Garden has for most people would not, I think, be tenable. This can hardly be the significance it has for atheists, after all. But the possibility is suggested of construing the account as testimony to a broader, but not theistically committed, conception of The Garden as epiphany—the kind, perhaps, which I develop later in this chapter.

A final strategy—definitive, perhaps, of a certain style of phenomenology—is to attend less to what people actually say about their understanding or attitudes, than to their practice. The strategy was at work, in effect, in Chapter 4 when considering the significance for people of particular garden-practices—a significance more apt to 'show up' through the ways they engage in these than in what they explicitly state. The premiss behind this general strategy, as we find it employed in the phenomenology of, say, Heidegger and Merleau-Ponty, is that, as the latter put it, understanding of or attunement to meaning is pre-eminently 'in the hands', not 'in the head'. Practice, dealing with things, Heidegger remarks, is not 'blind ... it has its own kind of sight' by which it is 'guided' (1980: 98), implicit as this typically remains for those whose practice it is. In the present context, the point is not about this or that garden-practice in particular, but about the understanding or sense implicit in people's more general engagement with The Garden. In a not unrelated context, when exploring the meaning that people may recognize their lives as having, John Cottingham puts the point in a helpful way: recognition is 'characteristically expressed through *practices* whose value and resonance cannot be exhausted by a cognitive analysis of [the] propositional contents' of people's explicit statements (2003: 99).

So the possibility arises for the enquirer of articulating the significance of The Garden—of identifying the X of which it is an epiphany—through attention to people's engagement with The Garden. This articulation will, of course, be an interpretation rather than reportage: but it need not be arbitrary or idiosyncratic, and will not be if it succeeds in enabling aspects of that engagement to 'fall into place', to be made sense of, and—to recall an earlier point—in 'ringing true' to those whose engagement it is. In what follows in this chapter, I shall, wherever appropriate, draw on this and other strategies in support of my attempt to articulate the meaning of The Garden.

A Modest Proposal

Indeed, let us straightaway deploy one of those strategies by drawing on the testimony of a man who, like his contemporary, Monet, was not only both painter and garden lover but, in his portraits of the gardener, Vallier, brought his two passions together. There are two themes which, for our purposes, stand out from Paul Cézanne's letters and conversations. The first is that of the artist's immense debt, too often unacknowledged, to nature. 'The artist must conform to [nature] ... Everything comes to us from nature; we exist through it' (in Merleau-Ponty 2004: 276); hence the artist—indeed, man himself—'has spent too long seeking himself in all he has done' (in Kendall 1994: 290). The second, converse theme is that of nature's debt to the human artist. While 'nature makes her meaning clear', she can do so only through the artist who co-operates to 'give' her that meaning, for it is not there just to take note of and copy (in Kendall 1994: 304). 'The landscape thinks itself in me and I am its consciousness', without which the landscape could not be exposed for what it is (in Merleau-Ponty 2004: 281). When the two themes are combined, the result is that art, the painter's or the gardener's, is 'the union of the world [nature] and the individual' (in Kendall 1994: 289). Cézanne, writes Merleau-Ponty (2004: 280), has in effect deepened 'the classical definition of art: man added to nature', by construing this adage in terms of the mutual debt of human beings and nature that is manifested in works of art.

I postpone the question of how Cézanne himself understood his remarks. Instead, I focus, in this section, on a relatively modest interpretation of what those remarks imply concerning The Garden as exemplification. My judgement is that The Garden does exemplify what, on the modest proposal, it is said to do, and it is important for this to be appreciated. The thrust of the final section of the chapter, however, is that, in exemplifying this, The Garden also embodies something further. The modest proposal, therefore, is over-modest.

According to the modest proposal, the import of Cézanne's first theme is that The Garden exemplifies the massive, but often unrecognized dependence of human creative activity upon the co-operation of the natural world. That of the second theme is The Garden's exemplifying the degree to which, more subtly, experience of the natural environment depends upon human creative activity. When combined, the two themes deliver the idea of The Garden as embodying a unity between human beings and the natural world, an intimate co-dependence. (The paintings of the gardener, Vallier, according to one writer on Cézanne, 'show humanity in harmony with nature' (Verdi 1992: 178).) This is not the unity or 'oneness' which some people claim to experience when communing with 'wilderness', not one that denies the distinctiveness of human beings from 'merely' natural beings. But in emphasizing both the dependence of human achievement on the assistance of nature and the dependence of our experience of nature on what we achieve, one deflates two views that threaten to render the natural environment alien or tangential to human endeavour. On one view, human creative endeavour is 'autonomous', dependent on little or nothing beyond people's own creativity and resolve. On the other, our experience of the natural world is dictated to us by an independent order, set over against us, and hence is not a function, to an interesting degree, of our engagement with the natural world.

There is little that is unfamiliar in what, on this proposal, The Garden is held to exemplify, which is one reason for describing it as 'modest'. Indeed, the claims just alluded to have been encountered, in various contexts, earlier in this book—in Chapter 2, for example, when indicating ways in which gardens differ from other artworks, and in Chapter 4 when discussing 'communion' with nature. Still, the proposal is a rich one which invites elaboration of aspects of the co-dependence of creative activity and nature that The Garden is held to exemplify.

It is sometimes argued that there is nothing distinctive about the garden's dependence on nature. It 'seems no different', two authors contend, 'from painting having to rely on the characteristics of oils or acrylics, [or] sculpture on stone or steel' (Kemal and Gaskell 1993: 20). They go on to argue that it is in environmental art (earthworks and the like), not gardening, that 'nature enters significantly into art', for here there is genuine 'interaction' between the two, not the 'supplanting' of a natural site by something man-made. We can ignore the jaundiced impression suggested by these words of gardeners, at large, simply 'supplanting' nature, as if replacing a meadow by a concrete car park, without any respect for the natural places in which they garden. The more important response is that, while painting and sculpture of course depend on natural materials, they do not typically 'refer to' or exemplify this dependence. Typically, the materials remain recessive: and even when attention is drawn to them, this is, with rare exceptions, only to what the work is made from, not to ways in which the work changes according to the vagaries of its materials. Indeed, painters and sculptors are generally at pains to insure against such changes. By contrast, gardens not only depend on nature but exemplify—'refer to', 'body forth'—this very dependence. It is in virtue of this, moreover, that gardens have a distinctive appeal for many people: it is a dependence both salient and significant for them.

There is truth, then, in Mara Miller's remark that gardens are 'more intimately and delicately dependent on the physical environment than any other art' (1998: 274). More even than architecture, she adds, for buildings and their appearances do not change in response to natural processes in as many ways, obvious or subtle, as do gardens. Obvious ways include the changes wrought in the garden by the seasons, the weather, the surrounding vegetation, plagues of insects, and so on. Subtler ways include the quality of the light and the position of the sun that may be crucial, as Gertrude Jekyll notes, to whether a gardener 'chanced to complete his intention' (1991: 61); or, as we saw Frederick Law Olmsted suggesting (p. 52), the 'states of the atmosphere, and circumstances that we cannot always detect, [which] affect all landscapes'. Moreover, The Garden, more palpably subject to 'misadventure and mischance' than The Building, more effectively explodes 'the myth of manageability' that infects our comportment towards the world—better exemplifies, that is, the precarious reliance on contingencies beyond 'the enforced will of human ingenuity' (Matthews 1990: 47 ff.). Nor, finally, should we ignore the kind of dependence on nature indicated by the *Sakuteiki*'s exhortation to 'follow the request of the stone' and other natural materials of the gardener's art (Takei and Keane 2001: 4). While the responsible architect will heed the natural environment in which his building is to be placed, the aesthetic quality and 'atmosphere' (see pp. 48ff) of the building is not dependent on detailed attention to the 'requests' made by the particular plants, stones, pools of water, and so on which occupy the site. Not dependent, that is, on a sensibility to the intrinsic properties of a flower, a stone's ability to combine with another stone, or the mirroring of the distant landscape by some bushes. Someone, certainly, should exhibit this sensibility, but this is the architect's colleague, the garden designer.

Turning to the second theme—The Garden as exemplifying the dependence of nature on human creative activity—no one will deny that many natural items in a garden, like flowers and vegetables, depend for their growth and flourishing on attention and tending by the gardener. Nor will anyone deny the obvious and gross physical changes in the natural environment that gardening, like building houses and roads, effects. These changes, moreover, are not confined to the immediate vicinity of the garden: re-routing a stream for irrigation purposes, or planting some bird-friendly trees, will impact on a wider ecosystem. But these are not the types of dependence hinted at by Cézanne's remark on the artist helping to make nature's meaning clear. This indicates, rather, the thought elegantly put by Robert Macfarlane (2004: 56) when he writes, 'in oil paintings of landscapes, the earth itself has been pressed into service to express itself'. He could as well have written this apropos gardens, and I want to distinguish three ways in which the thought might be elaborated.

In Chapter 3, during my criticism of the 'factorizing' approach to the aesthetic appreciation of gardens, I argued that such appreciation could hardly be a function of nature appreciation in cultures where the latter is undeveloped or in those where nature appreciation has been significantly shaped by that of gardens. In the present context, the point is simply to emphasize that experience of 'wild' nature has indeed been shaped by garden traditions, as it has by those of other arts. There is no need, without severe qualification, to accept Arnold Berleant's conclusion that nature is a 'cultural artefact' in order to subscribe to his premiss that our 'very conception of nature has emerged historically, differing widely from one cultural tradition to another' (Berleant 1993: 234). It seems, for instance, to have been the purpose of Italian Renaissance designers, in placing their formal, manicured gardens on hilltops overlooking great landscapes, to induce a heightened sense of the rudeness of uncultivated nature. And, whatever their intention, makers of 'picturesque' gardens in the eighteenth century encouraged a Romantic conception of nature more applicable to these gardens than to the natural places which the gardens were supposed to picture.

Another criticism of the 'factorizing' approach, illustrated by Wallace Stevens's 'Anecdote of the Jar' (p. 57), was that artefactual and natural elements of a garden are too 'holistically' interwoven for appreciation of the garden to be the product of their separate appreciation. The point, in the present context, is to emphasize the degree to which experience of natural features in the garden or its vicinity is dependent on human artistry. This is akin to the point made by Heidegger in a celebrated passage where he describes the impact on the Greeks' experience of the natural world of a temple built in a 'rock-cleft valley'. It is the temple that 'first gives to things their look'—trees, grass, sea, animals, birds, plants—and enables these things, and earth itself, properly to 'emerge' for the Greeks. We must recognize, therefore, 'how differently everything then faces us' in nature once we have placed our creations in its midst (Heidegger 1975: 42 f.). Heidegger could have taken certain gardens instead of a building to make his point: elsewhere, indeed, he employs the German verb *bauen* ('to build'), in keeping with its etymology, to refer to cultivating as well as building. He might, say, have spoken of the gardens designed by Geoffrey Bawa in Sri Lanka, created in a jungle clearing or perched above the Indian Ocean. Doubtless, Heidegger makes his point in exaggerated terms: neither the Greek temple nor the garden at Lunuganga literally first give their 'look' to olive- or coconut-trees, to eagles or egrets. But it would be hard to deny that a garden, perhaps through its harmonization with a surrounding landscape, perhaps through its contrast with it, may render salient to experience significant and otherwise unremarked aspects or 'moods' of this landscape. The very artificiality of the buildings and gardens of Isola Bella, on Lake Maggiore, argues Edith Wharton, induces an experience of the lake's scenery, in contrast to the rougher Roman landscape, as one which 'appears to have been designed by a lingering and fastidious hand, bent on eliminating every crudeness and harshness' (Wharton 1988: 206).

Plate 7. Garden, buildings, and scenery, Isola Bella, Lake Maggiore, Italy

Finally, there is a way in which gardens may help nature to 'express itself' suggested by Heraclitus's pronouncement that 'nature likes to hide'. Probably he meant something more profound than that there are aspects of familiar natural phenomena which ordinary experience, without quite distorting the phenomena, nevertheless typically fails to expose, to make available to thought. Profound or not, the point is important, and so, therefore, is the fact that gardeners often facilitate, through devices that impinge on the senses, the exposure or 'expression' of these aspects. Let me give two examples of this kind of dependence on gardens of our experience of nature. 'The Japanese garden designer', wrote Ezra Pound, 'creates a theatre for the wind to speak' (in Keane 1996: p. xii). The poet's point could be put by saying that the Japanese garden, through its bamboos and wind-chimes, expands the range of meanings that the wind may have for us. Blowing through the garden, it is not the enemy we encounter in our face on the way to the office, nor the friend that cools us on a baking beach: rather, we experience it as animator. In 'speaking' through the chimes and bamboos, it animates certain qualities of these: those qualities, as it were, themselves come to speak—the resonance of the wood from which the chimes are made as they knock against each other, the delicate but sturdy flexibility of the bamboos as the wind rustles and bends them. (Pound's remark and perhaps the practice of the garden designers he refers to are inspired, one surmises, by the famous passage on the 'piping' of the earth and of heaven in Chapter 2 of the *Chuang-Tzu* (see Graham 2001: 48 f.).)

To take another example, I find that my image of winter is less than it once was the image of a time of bleakness and barrenness, less that of the dead season that Vivaldi's 'Winter' so effectively conjures for us. This may be because I now spend quite a lot of it abroad or cocooned in centrally heated rooms, but I suspect that it is also due to the increasing tendency of gardens, my own included, to display the colours of winter-growing flowers—hellebores, penstemons, and others. What the display of colour vividly reveals is that, as Čapek reminds us (2003: 152), vegetation does not stop in November: rather 'it has rolled up its sleeves', and beneath the flowers there is a world of 'secret bustling'.

The 'modest proposal' is an attractive one. The two thoughts fused in the claim that The Garden exemplifies the co-dependence of human creative activity and nature are persuasive. At least, I hope they are, for unless they actually persuade, then, as explained in the previous section, this would be a reason to reject them as truths about the meaning of The Garden. This meaning must be 'available', something people can acknowledge. Canvassing for such acknowledgement, it was also explained, is not the only strategy to deploy in support of a claim about The Garden's meaning. The testimony of people who reflect on and articulate what The Garden exemplifies is also relevant, as, more crucially, is the understanding implicit in the practice of those to whom The Garden matters. On both these counts, I suggest, the modest proposal does well. I have assembled some relevant testimonies over the last few pages, and there are many more which might have been. And it is surely the case that many people, whether or not they articulate their understanding in the vocabulary of co-dependence, exhibit understanding of this in their practices. For example, people design their gardens, and comport themselves within them, in ways that exploit the changes, physical and perceptual, induced by the seasons or alterations in the light. Gardens get situated, and admired, for the impact they have on experience of the surrounding environment. Again, various garden-practices, to recall Cottingham's words, have an acknowledged 'value and resonance' for people, irrespective of their willingness or ability to speak of this in 'propositional' terms, in virtue of the kind of 'communing with nature', identified in Chapter 4, which these practices enable.

And yet, I want to say, the modest proposal, despite its success, is over-modest. The Garden, in exemplifying the co-dependence of nature and creative endeavour, embodies—is an epiphany of— something further. What this might be is the topic for the next section.

A Further Proposal

Cézanne would not have been satisfied with the modest proposal's construal of his remarks. When he observes that 'nature is always the same' and that art must provide a 'sense of her permanence', her 'eternal qualities', that belies 'the appearance of her changeability', it is apparent that the nature on which the artist depends is not the natural world at all, neither the one we familiarly perceive, nor the one described by science. 'What I am trying to convey through art', he adds, 'is more mysterious' than the natural world and our relationship to it: 'it is bound up with the very roots of being, the intangible source of sensation'. We do not go far enough, he implies, in regarding painting or The Garden as an epiphany of that relationship, since nature itself is a 'show', spread before us by *'Pater Omnipotens Aeterne Deus'*, a 'catechism' of its 'author' (in Kendall 1994: 236, 293, 302 f.). If The Garden exemplifies or embodies co-dependence, then, this cannot simply be that between human endeavour and nature, but a further, 'more mysterious' relation.

But why should we follow Cézanne here and suppose that the meaning of The Garden is more mysterious than it is on the modest proposal? Isn't what he finds in The Garden a merely 'personal' meaning, not *the* meaning? It would be quite wrong, however, to suppose that it is 'personal' in the sense of being idiosyncratic or eccentric. Indeed, a main reason to remain dissatisfied with the modest proposal is the massive documentation of a sense of The Garden's meaning that is at least akin to the one expressed by Cézanne. We encountered some of this documentation, and an important aspect of the sense to which it attests, in Chapter 4, when Michael Pollan spoke of his Sibley squash as a 'gift'. His point was not the modest one that success in growing the vegetable owed to natural processes beyond his control, but that these very processes were experienced as if they were 'given' by something—grace, as it were. And we encountered more documentation earlier in the present chapter when recording some of the theistically charged references, which abound in the literature, to The Garden as an epiphany of a creator. (To judge from his mention of the 'author' of nature, *Pater omnipotens*, it is to this tradition that Cézanne himself belongs.) Those references, I argued, cannot supply an account of The Garden's meaning that could be acknowledged at all generally, since they invoke a doctrinal belief, in a creator-god, which is clearly rejected by many people to whom The Garden matters. But it would be possible, I suggested, to regard those references as testifying to a wider sense of The Garden's meaning that need not take a specifically theistic form.

It is important, certainly, to appreciate that there are traditions in which the sense of the significance of The Garden, while inviting the label 'spiritual', even 'religious', is free from any doctrinal commitment to theism. Writing of the Japanese art of *ikebana*, Gustie Herrigel (1999: 119) proposes that while 'there are many things in flower-setting that can be said . . . yet behind everything that can be . . . represented there stands, waiting to be experienced by everyone, the mystery and deep ground of existence'. Another author, with Japanese gardening on a rather larger scale in mind, speaks of it embodying a 'presence of transcendence' that 'infuses' human activity with a sense of 'governing spiritual purposes' (Matthews 1990: 51). Although such remarks are by no means confined to the literature of Zen Buddhist gardening arts, these are arts which have indeed been practised in the spirit and sense captured by the remarks. In Zen tradition, Yuriko Saito observes, gardens are held to afford an especially revealing 'glimpse of this world as it appears to a Zen-enlightened sensibility' (in Carlson 2000: 173 n. 24).

The word 'sensibility' in Yuriko Saito's remark needs emphasizing. Zen is not a theistic dispensation: indeed, the word 'doctrine', even 'belief', is an inept one for capturing Zen's attunement—its sensibility—to a spiritual or mysterious 'ground' of our ordinary, worldly existence. It is a sensibility that, as Herrigel put it, might be 'experienced by everyone', irrespective of doctrines and belief, theistic or otherwise. (Nor, of course, is the point confined to experience of Japanese gardens alone.) It is not a sensibility, therefore, which could be acknowledged only by people with a doctrinal commitment to a creator-god—and not one, therefore, from which people with other doctrinal commitments, to atheism for example, are necessarily excluded. Maybe, that is, it is a sensibility and understanding that everyone with a sense for the meaning of The Garden could acknowledge and own to. For it is a sensibility exposed, not through 'cognitive analysis of [the] propositional contents' of people's assertions and beliefs, but through reflection on their engagement with The Garden. My following remarks are an attempt at such an exposure: readers will judge how successfully or otherwise I articulate their own sensibility, their own sense of The Garden's meaning. (I do not try to defend the kind of sensibility that, here, I am aiming only to expose, albeit sympathetically: but I do elsewhere (Cooper 2002), though not with special reference to gardens.)

According to the modest proposal, The Garden exemplifies a co-dependence between human endeavour and the natural world. On my present, as it were immodest, proposal, this co-dependence itself embodies or refers us to the co-dependence of human existence and the 'deep ground' of the world and ourselves. By embodying something that itself embodies something further, The Garden—in one of Goodman's 'chains of reference'—embodies this 'something further'. The Garden, to put it portentously, is an epiphany of man's relationship to mystery. This relationship is its meaning.

Portentous or not, the thought that human existence depends on something mysterious is not especially opaque. We are prone to think of our achievements as ones of our own making and doing. At an obvious level, we are right to think this: it was I who wrote this book, my friend who built that greenhouse, Cézanne who painted that landscape. But we are also prone, thereby, to ignore the pre-conditions—not at all of our own making and doing—of these achievements. For anyone to do anything, there must already be, one might say, a space of possibilities. And for there to be this space, there must already be some general understanding of the world and ourselves; already a sense of what matters, of what would be worth doing; already available 'moods' and 'attunements' that enable aspects of our world to assume a certain tone, attractive, repellent, or whatever; already be a light in which things show up for us in the ways they do and invite us to treat them in this or that manner. None of what must already be in place is our own doing, but is rather the condition of our doing anything. Nor can what is responsible for all of this being in place be described or conceptualized: for the descriptions we give and the concepts we forge—these are our achievements. Like any other achievement—this book, my friend's greenhouse, Cézanne's picture—they presuppose what is not of our making. This cannot be described or conceptualized, since it is the condition for the possibility of any descriptive terms or concepts with which we manage to equip ourselves.

Human creativity, then, must be what the French philosopher Gabriel Marcel calls 'creative receptivity'. Even the most Promethean artist or inventor should renounce 'the claim that ... we have the power to make' things 'dependent only on ourselves'. Instead, we should cultivate 'wonder' at what is 'granted to us as a gift', for we at most 'welcome' and 'transmute' what, by no 'device' of our own, comes to presence for us. Marcel's name for what grants the gift is not 'nature', but 'Being' (Marcel 2001: i. 208; ii. 32, 87 f.).

Such, then, is the thought that inspires one half of the immodest proposal. In its obvious dependence on the co-operation of the natural world, the endeavour of the gardener embodies 'creative receptivity', the dependence of our doings and makings on what is gifted to us. But only one half—for it is *co*-dependence that, according to my proposal, The Garden embodies. What, then, may be said of the seemingly opaque idea that the gift, to stay with Marcel's term, is dependent on us and our endeavours? Something mysterious may be the condition of the latter, but how can they be a condition of it—their alleged 'ground'?

To see how, recall some aspects of the dependence of nature on human practice discussed in the previous section: the manner in which human experience of the natural world—sea, sky, animals— is shaped by such creations as buildings, farms, and gardens; and the way that otherwise unremarked dimensions of natural phenomena, like the wind, are exposed through human artistry. Now the general point indicated here is not confined to experience of the natural world, or to the effect of particular creations. The general point is that all human experience of the world crucially owes something to the ways in which we engage with it. But this point yields a more radical result when we consider that there is no final distinction between how, due to this engagement, we experience the world and how the world is. If *any* account we give of the world reflects our engagement with it—our purposes, interests, sense of what matters, and so on—then there could be no world for a creature without any such engagement. Its experience would not be of a structured, articulated order, for the structure and articulation we discern are a result of our engagement. In other words, this creature would not experience a world at all, but at best what Nietzsche called 'a chaos of sensations'. So the more radical point, in Heidegger's idiom, is that human engagement, in being necessary to our experience of the world, thereby 'lets the world be as world'.

The world, then, is dependent on us, on our engagement and practice. But how do we move from this to the claim that the 'ground' or 'gifting' of the world 'needs [man's] presence' (Heidegger 1975: 228)? Peculiar as that claim may sound, it is central to the several religious traditions which urge that this 'ground' is not disjoined from the world, in the manner, say, of a creator-god. The claim is there, for example, in the Zen perception that the 'emptiness (*śūnyatā*)' to which the ordinary, experienced world owes its existence is not an entity distinct from the latter: indeed, they are spoken of as being 'non-different'. Such utterances may do little to satisfy someone who finds the claim peculiar to the point of unintelligibility. But then 'claim', like 'doctrine' or 'belief', is not really the apt word for what is being voiced. Rather, here at the limits of language, one should hear utterances like 'Being needs man's presence' as gesturing at, attuning to, a certain sensibility—the 'Zen-enlightened sensibility' mentioned earlier (albeit one not confined to practitioners of Zen). Such a sensibility, Herrigel observed, is 'waiting to be experienced by everyone', and I suggest, to borrow a favourite Zen phrase, that it is indeed 'nothing special', not the preserve, certainly, of a few adepts. (On Zen sensibility, see James 2004.)

I want to urge, predictably, that a garden is a place especially conducive to this sensibility. The garden, certainly, rather than the study: for one reason why it is 'nothing special' is that it is manifested, not during one's armchair metaphysical musings, but in the midst of ordinary activities—'carrying water and logging firewood', according to one Zen poet. In this respect, it is akin to the gardener's understanding, which, as one distinguished gardener records, only 'physical contact' can give—an understanding more 'through the hands [than] through the head' (Page 1995: 16 f.). To become sensible to a mysterious 'power', wrote the great Zen master Dōgen, nothing 'gigantic' has to be experienced: a cypress-tree or the moon reflected in one's pond will do (1996: 75).

Sudden and poignant, or steady and undramatic, the sensibility— or mood, or attunement—I have in mind is present when, first, one enjoys, with an 'innocent eye', as Dōgen puts it, a sense of experiencing things 'just as they are', a sense that 'nothing has been hidden' from one (1996: 3). The wood one chops, the squash one waters, the aubretia one secures to the stone wall—each is there vividly and saliently before one 'just as it is'. But this is not at all the sense, secondly, which the detached scientist or botanist might have when subjecting the same things to objective inspection and analysis. On the contrary, it is at the same time a sense of intimacy with them, an engaged experience of them: of the squash as something that needs to be watered, of the aubretia as gracing the wall that protects the cottage garden. To such a sensibility, then, things are present 'just as they are', not *despite* the place they have in relation to our lives, but *through* this. *This* squash, *this* aubretia—these are not simply bits of matter with certain shapes and botanical properties, for they are what they are through the particular ways they engage the gardener. (One is reminded of Heidegger's point, in *Being and Time*, that it is in and through hammering that the hammer is understood for what it is: a hammer, and not simply a shaft of wood attached to a lump of metal.) Indeed, to abstract these things from their context of engagement—being tended and grown for food, or planted to punctuate a wall with colour—is to cease experiencing them as vegetables or flowers at all.

To experience the world as intimate with ourselves, and yet with a sense that this is the world 'just as it is', is to recognize it as 'our world'—not one behind which there lurks, hidden or behind a 'veil of perception', a world discernible only by a mind detached from all purposive and affective engagement with things. It is the world itself, not some surrogate for it, which presents itself to 'the innocent eye'. None of this means, however, that for the sensibility I am describing the world and its ingredients—the squash, the aubretia— are regarded as our products, spun out of our minds, the arbitrary constructs of some 'conceptual scheme' we have 'chosen'. The world is 'our world', but not thereby one 'the law of [whose] making' is of our promulgation (Merleau-Ponty 2002: p. xxi). On the contrary, for an 'enlightened' sensibility, things present themselves as given to us. This cannot be, merely, the appreciation that the squash or aubretia owes to natural phenomena—the rain, the rich soil—which are not of our making. For the rain and the soil are themselves experienced, in the relevant mood, as intimate with ourselves. They, too, belong to 'our world', which, as a whole, arises for or, as Dōgen expresses it, 'advances' towards, us. There can be no literal statement of the ground or 'intangible source', as Cézanne called it, of this 'advance': only wonder, muted and gentle though this may be, at its mystery—at the mystery of the arising of a world with which we are intimate, at 'something ineffable coming [to us] like *this*' (Dōgen 1996: 3).

For the sensibility in question—'Zen-enlightened' but 'waiting to be experienced by everyone'—the world is 'our world' and so 'needs [man's] presence', but it is nevertheless a world simultaneously experienced as a mysterious gift. It is a sensibility, I suggested, peculiarly apt to be summoned by our engagement with gardens: for salient in gardening and other garden-practices is a co-dependence between our creative activity and the natural conditions for this activity. This is a co-dependence people's appreciation of which may modulate, in the appropriate mood and when appropriately attuned, into a sensibility to the world as a gift that 'needs' us, its creative recipients. It is true, of course, that this is a sensibility that it is *possible* for people to have in relation to any creative activity. The artists whom Marcel enjoined to renounce the conceit that their creative power depends only on themselves are *all* artists, not just garden designers and gardeners. But if we are seeking to identify a creative activity that pre-eminently invites such a renunciation and invokes the sensibility in question, it is surely that of the person co-operating with the gifts of nature in making or tending a flower-bed or a rock garden, not that of someone in a studio smearing man-made paints on a man-made canvas. This remains so even when the painting's 'message'—not an unfamiliar one in 'postmodernist' art—is that the artist is not the 'autonomous', 'controlling' subject that he or she was taken by 'Modernists' to be. I would suggest, as well, that gardening more effectively evokes the sensibility in question than works by 'land' and 'body' artists which are transparently contrived to hammer home the thought that human creations are heavily dependent on the co-operation of nature. (One thinks, for example, of Ana Mendieta's 1970s series, *Silueta*, in which, for instance, the imprint of her body in the snow is transformed as the snow thaws. (See Perry 2003).)

The Garden, then, is an epiphany—a symbol, in the Romantic sense—of the relation between the source of the world and ourselves. It is, in the terminology of Chapter 6, peculiarly 'appropriate' to this relation, which is why, over the centuries, paintings of or poems about gardens have exploited the symbolism of the garden and themselves served as vehicles for evoking the sensibility of which I have spoken. And it is the reason, more importantly, why I am also suggesting that in this epiphany the meaning of The Garden, its deep significance for people, is located.

How does this suggestion bear upon my discussion, in Chapter 5, of gardens and the good life, and indeed upon other discussions in earlier chapters? In the next and final chapter, I shall make these connections.

Coda

Before that final chapter, however, I want to add a few words on a type of garden that my argument in the present one makes it virtually imperative to discuss, partly because it might seem to pose a problem for that argument.

I have referred to the sensibility which engagement with 'exemplary' gardens is especially apt to summon as 'Zen-enlightened'. It would be unfortunate, therefore, if the types of garden particularly associated with Zen practices turned out not to be 'exemplary', not to be ones apt to symbolize the co-dependence of world and human creative activity of which I spoke. If Zen gardens offer a 'glimpse of this world' as it is understood in Zen thought—if, as the Japanese philosopher Nishitani Keiji puts it, 'the garden is my Zen master' (in Berthier 2000: 136)—then they should be conducive to an experience of co-dependence. Otherwise my reference to a Zen-enlightened sensibility was unwarranted.

One type of garden particularly associated with Zen is the Japanese 'dry landscape' (*karesansui*), such as those at the Ryōanji and Daitokuji temples in Kyoto. Composed almost entirely of rock, gravel, and sand, and almost devoid, therefore, of vegetation, such gardens have been held by some not to deserve the title 'garden' at all: but it is surely more plausible to count them, as I did in Chapter 1, as 'atypical' gardens. (They are 'atypical' not only because of their lack of flowers, trees, and so on, but because they are designed purely to look at or contemplate, not to move about in.)

It is not my purpose here to offer a full account of the power and fascination that Japanese dry gardens clearly exert on visitors, of the many ways in which, as Graham Parkes puts it, they 'speak to us'. (See his 'Philosophical Essay' in Berthier 2000.) Any such account would need to discuss the associations, actual or fancied, between these gardens and elements in Shinto, Daoist, and Buddhist mythology, as well as the reverence paid to rocks in East Asian traditions to which a sharp distinction between the animate and the inanimate in

Plate 8. Dry landscape, Daitokuji Temple, Kyoto, Japan. Photograph by Bret
Wallach, reproduced by permission of greatmirror.com

a world perceived to be pervasively charged with 'psychophysical
energy' (Chinese *qi*) is foreign.

My purpose, rather, is to indicate and then respond to a problem
for my own construal of 'Zen-enlightened sensibility' that *karesansui*
might seem to pose. An important reason, I argued, why exemplary
gardens are an appropriate epiphany of co-dependence is the
ways—rehearsed in the section, 'A Modest Proposal'—in which
gardens depend, however great the human endeavour that goes into
their creation and maintenance, on the co-operation of the natural
world and its processes. Now there are clearly ways in which the
dry garden is less dependent on this co-operation than the 'typical'
garden, with its flowers, shrubs, grass, and the like. It is not markedly
subject to processes of growth and decay, to the vicissitudes of the
seasons, or the impact of animal and insect life. More generally, such
a garden—stripped of vegetation and water—seems, like 'the bones

of a skeleton', to 'defy time' (Berthier 2000: 32). As such, how can the dry garden serve as an apt symbol, an epiphany, of co-dependence? Once set up by its designer through having rocks hauled into place and gravel sprinkled, and with only minimal maintenance then required, the garden appears to continue on its own immutable course. The garden, neither in its setting up nor its maintenance, calls for much co-operation on nature's part.

This line of thought, I suggest, is insufficient to disqualify the dry garden from being an epiphany of co-dependence. This is so even when, as in the preceding paragraph, the focus is on what, on p. 137, I called the 'obvious' ways in which gardens typically depend on the co-operation of nature. It is worth noting that the dry gardens at Ryōanji and Daitokuji are not entirely stripped of vegetation: they contain moss or small shrubs which, despite, or rather because of, their modest but unmistakable presence, remind the viewer that here are places where natural processes are still at work. More importantly, dry gardens are typically set against or within surroundings of trees or other plantings—subject to obvious processes of growth, seasonal change, and so on—that serve as 'borrowed scenery' (*shakkei*) which enters crucially into the experience of the gardens.

But there were, of course, 'subtler' ways in which, I urged (p. 137), gardens depend as objects of experience on the co-operation of nature—on effects of light, say, or on sudden showers. And to these effects of nature, the dry garden is as much, or even more, subject than more 'typical' ones. The sun's 'rays filter through the trees and highlight different elements of the composition differentially. When the [nearby] branches sway in the wind, light and shade play slowly over the entire scene, the movements accentuating . . . the stillness of the rocks' (Parkes, in Berthier 2000: 86). And there is one 'subtle' form of dependence on nature that is palpably manifested by the Japanese dry garden—the one, indeed, which I illustrated by the *Sakuteiki*'s exhortation to 'follow the request of the stone'. This is the dependence of a successful garden creation on the intrinsic properties of materials—rocks, say—and the designer's proper attention to their 'demands'. It has often been remarked of the composition of fifteen rocks at Ryōanji that it is as if each rock *has* to be placed where it is, as if the rocks, in virtue of their own qualities and their relation to one another, demand so to be placed.

Finally, it is an error to suppose that the seemingly immutable, time-defying character of the *karesansui* disqualifies it as a symbol of co-dependence. The bones of a skeleton, to recall Berthier's analogy, may change little, but, precisely for that reason, be an effective reminder of change, of the processes of life and nature that have now ossified. (Yorick's skull reminds Hamlet of his boyhood more effectively than the head of his friend, when still living, might have done.) Likewise, a dry 'waterfall' or 'lake' in a Japanese garden may be more evocative of water's role in our lives and our dependence upon it than the real stuff is. More generally, the very absence of the natural processes on which our activities, gardening included, depend, may, when suitably evoked, become more present to our minds than when they are there visibly before us. With only seeming paradox, therefore, the dry garden can, through its very immunity to certain natural processes, render our ordinary dependence upon these all the more palpable. And that is why such gardens can be epiphanies of co-dependence between human creativity and that 'out of which "all things arise"' (Parkes, in Berthier 2000: 109). And that, in turn, is how these gardens do, after all, summon the 'Zen-enlightened sensibility' I described.

8   CONCLUSION: THE GARDEN'S DISTINCTION

What musicologists call a leitmotiv of a piece does not need to blaringly obvious, but once it is identified, the piece may then be heard as a more organized whole than it sounded at first. An unobtrusive leitmotiv of this book has been the distinctiveness of the garden—or, better, of The Garden, to recall my orthographical device for speaking about 'exemplary gardens' (p. 123). This theme first appeared in the context of rejecting 'assimilationist' accounts of the (broadly) aesthetic appreciation of gardens. By the end of Chapter 3, I had concluded that such appreciation cannot be reduced to that of (other) arts or of 'wild' nature, or to a combination of these types of appreciation. There is, as one author put it, a '*unique* strength' to gardens as objects of appreciation. The theme re-emerged in Chapter 4, this time in the context of arguing for the distinctiveness of 'garden-practices'. These were not practices that, as it happens, can be conducted in, amongst other places, the garden; for, once conducted there, they assume a special 'tone', as I put it. Expressed with pardonable paradox, they cease to be the same practices that they are when conducted outside the garden. The theme was sounded again, in Chapter 5, where it was argued that various practices to which gardens are especially hospitable 'induce' a number of virtues, including care, humility, and hope. As such, The Garden makes a distinctive contribution to 'the good life'. Finally, the theme was heard yet again in Chapters 6 and 7, during my discussion of gardens and their meanings. While gardens, like much else, have meanings—expressive, symptomatic, depictive, and so on—in many different modes, I argued that there is such a thing as 'the meaning of The Garden'. As an 'epiphany' of a certain relation between human

creative activity and 'the mysterious ground' of the world in which
human beings act, The Garden is peculiarly 'appropriate' to this
relation of 'co-dependence': it has this relation as its meaning.

My claim about The Garden as epiphany was, to pursue the
musical metaphor, the climax of this theme, and, like some climaxes
in music, it enables the theme in its earlier appearances, when these
are revisited, to sound out with greater resonance. We can now
surmise, for a start, that the distinctiveness, the irreducibility, of
appreciation of exemplary gardens and their 'atmosphere' is due,
on some occasions at least, to a sensibility towards these gardens as
epiphanies of co-dependence—something that 'wild' nature and
other art-forms are not. Again, we may surmise that the distinctive
'tone' of some garden-practices—growing one's own food, say, or
reverie (pp. 83f)—owe to that same sensibility. Finally, and perhaps
most importantly, given that my 'fundamental question' was that of
why gardens 'matter', we may revisit the issue of The Garden's contribution to the good life, now in the light of the idea of The Garden
as epiphany. This is what I shall be doing in the remainder of this
concluding chapter.

My discussion of this contribution, in Chapter 5, ended in effect
with a question. Gardens contribute to the good life, I argued,
through being 'hospitable' to certain practices that 'induce' virtues,
in the sense that these virtues are 'internal' to such practices when
these are carried out with proper understanding. The squash grower,
for example, who fails to exhibit care towards his vegetables is failing
to understand what the practice requires. But the question then
arose: What makes these virtues virtues?, or, which amounts to the
same thing, How do these virtues relate to the good life? Two broad
answers to this question were canvassed. According to one answer,
these alleged virtues contribute to the good life through being conducive to happiness; according to the other, it is because they are
ingredients of a life that is 'in the truth', a life which manifests recognition of the place of human existence in the way of things (p. 98).

We can now ask how the idea of The Garden as an epiphany of
co-dependence bears upon the above question and the answers to it

which were canvassed. If, as I think, the notion of co-dependence incorporates an important truth about human existence and the world, and if The Garden does, as I argued, embody or exemplify co-dependence, then engagement with The Garden is 'in the truth'. It is, at any rate, when informed by an appropriate sensibility towards what The Garden embodies. Given this, it will follow that the virtues 'induced' by garden-practices indeed contribute to the good life if the latter is characterized as a life that is led 'in the truth'. We should, that is, see qualities such as care, humility, and hope as aspects of a life informed, however implicitly, by a sensibility towards a fundamental truth of the human condition.

The point becomes clearer if we recall a generalization I made at the end of the discussion of the virtues in Chapter 5, where I suggested that the virtues identified—and perhaps all virtues—belong to the wider economy of 'unselfing', as Iris Murdoch called it. Now the person with a sensibility towards co-dependence appreciates, in effect, that human creative activity is both dependent on what is 'beyond the human' and, at the same time, essential for a world to arise at all. But a life which manifests appreciation of these truths is one which must also manifest humility—a denial of the 'autonomy' of human agency—and a sense of care and responsibility for what is 'gifted' to us. In both respects, it is an 'unselfed' life. The person, that is, who engages, suitably attuned, with The Garden displays—at least in this sphere of his or her life—what is called for by appreciation of the place of human beings in the way of things.

Does this mean that the virtues of The Garden have only to do with truth, and nothing, therefore, with happiness, the other traditional criterion of the good life? I argued in Chapter 5 that happiness cannot be the criterion when it is understood in the modern sense, in terms primarily of pleasure and mere contentment. But that is not how happiness has always been understood—or, to be more exact, not how certain words like *eudaimonia*, which have been questionably translated as 'happiness', were originally understood. In particular, it was once understood, in Greece and India alike, pre-eminently in terms of the serenity, peace of mind, and satisfaction that come with a sense of living 'in the truth', and thereby themselves become ingredients of the good life. This is why, for example, the Stoic, Seneca, could regard *gaudium sapienti*, 'the happiness of the wise', as 'the greatest good' (1917–25: lix. 14). Happiness and truth, then, need not be rival or mutually exclusive answers to the question concerning the contribution of the virtues to the good life.

Garden literature, as noted earlier, is predictably replete with lyrical tributes, like Andrew Marvell's, to the satisfactions, serenity, and mental peace that gardens and gardening afford. Sometimes the point intended by such tributes barely extends beyond cataloguing the benefits of fresh air and exercise, and the catalogues usually omit the back-aches, the other pains, and the frustrations to which gardeners are prone. But there is a deeper, more valid point to be made. For all the aches and frustrations, the dominant 'tone' of enlightened engagement with The Garden will be the 'happy, calm, unshaken' one spoken of by Seneca (himself a garden lover)—for this is the 'tone' of practices pursued in the recognition that they belong to the good life.

There is a poem by Martin Heidegger, called 'Cézanne', which besides pleasingly looping us back to my discussion of the French painter, encapsulates much of what I am trying to articulate in this and the previous chapter. Here is Julian Young's translation of the relevant section of the poem:

> The thoughtfully serene [*Gelassene*], the urgent
> Stillness of the form of the old gardener,
> Vallier, who tends the inconspicuous on the
> *Chemin des Lauves*.

And here is Young's gloss on these lines: 'The old gardener's "tending" is his passive caring-for the earth. And his "urgent stillness" is, I suggest, an action-ready *listening*—a listening for and to "the request made by the earth"' (Young 2002: 108; the last phrase is from the *Sakuteiki*). Heidegger's point, so construed, is that there is serenity, and therefore happiness, in caring for living things in response to their needs and demands. I suggest, however, that Heidegger's

Plate 9. Paul Cézanne, *The Gardener Vallier,* c.1906, Tate Gallery, London. Reproduced by permission of the Tate Gallery.

thought extends well beyond this point. The ordinary German word *Gelassenheit* indeed refers to serenity, and Heidegger does not want us to ignore that meaning. But in his writings, the expression also has the sense of 'releasing' or 'letting be'. A person who is *gelassen* is not only serene, but is letting something be, appear, become present; indeed, the reason, for Heidegger, that this person enjoys such serenity is that he or she is not imposing, not dictating how things shall be. Similarly, the word 'inconspicuous', in Heidegger's later writings, indicates something other than being hidden or obscured in the obvious way that features of the earth or soil, to which the gardener must attend, may be. In fact, he uses the word to describe the very 'ground' of all things, of the world: something that is not itself noticed, but is the condition for there being anything to notice. The serene gardener who 'releases' and cares for the products of the earth, therefore, exemplifies or embodies the relation of co-dependence: the relation, in Heidegger's vocabulary, between a human agency that 'releases' things and the inconspicuous 'ground' of the world that presences for human beings.

Heidegger's poem should be taken in conjunction with a remarkable lecture he gave in 1951, 'Building Dwelling Thinking'. Building, in Heidegger's sense, includes, as I noted earlier, cultivation. And cultivation, in his sense, is less the 'making [of] anything', than a caring for, a 'preserving and nurturing', as when the gardener 'tends the growth that ripens into its fruit of its own accord'. Heidegger argues that we do not, as conventionally assumed, 'dwell because we have built [and cultivated]'; rather, 'we build [and cultivate]... because we dwell'. 'Dwelling' is his name for what is, or at any rate should be, 'the basic character of man's being'. An enlightened or 'authentic' human life is that of 'dwelling'. Clearly, by 'dwelling', Heidegger does not only mean living somewhere, rather than nowhere: to dwell is to live somewhere, but in a certain way. What this way is, is suggested by the etymology of the German word *wohnen*, which Heidegger traces back to words meaning peace, freedom, sparing, and preserving. In effect, to dwell is to 'remain at peace' through freeing or sparing, and then caring for and preserving, things (Heidegger 1975: 146–50). To free things is to allow them to be experienced as the 'gifts' they are, to allow the world to become present for us through our engagement with it, but without our imposing upon them alien purposes. In other words, the authentic dweller is *gelassen*: he or she serenely 'lets be'.

The poem and the lecture, then, combine to present an image of gardening or cultivation as a practice which, engaged in with an appropriate sensibility—engaged in 'thinkingly', as Heidegger would say—embodies more saliently than any other practice the truth of the relation between human beings, their world, and the 'ground' from which the 'gift' of this world comes. In Heidegger's poem, as in Cézanne's paintings of the same man, the gardener Vallier becomes the peculiarly appropriate embodiment or symbol of a serene life led in attunement to truth. People who are able to recognize him as this symbol are on their way to being able, as well, to recognize why The Garden is distinctive and why The Garden matters.

REFERENCES

Adams, William H. (1991) *Gardens through History: Nature Perfected*. New York: Abbeville.

Adorno, Theodor (1997) *Aesthetic Theory*, trans. R. Hullet-Kentor. London: Continuum.

Allison, Henry E. (2001) *Kant's Theory of Taste: A Reading of the* Critique of Aesthetic Judgement. Cambridge: Cambridge University Press.

Andrews, Malcolm (1999) *Landscape and Modern Art*. Oxford: Oxford University Press.

Annas, Julia (1993) *The Morality of Happiness*. Oxford: Oxford University Press.

Appleton, Jay (1975) *The Experience of Landscape*. New York: Wiley.

Aristotle (1999) *Nicomachean Ethics*, trans. T. Irwin. Indianapolis: Hackett.

Bachelard, Gaston (1994) *The Poetics of Space*, trans. M. Jolas. Boston: Beacon Press.

Bacon, Francis (1902) 'Of Gardens'. In *Essays*. London: Grant Richards, 127–33.

Berleant, Arnold (1993) 'The Aesthetics of Art and Nature'. In Kemal and Gaskell (1993), 228–43.

Berthier, François (2000) *Reading Zen in the Rocks*, trans. and with a philosophical essay by G. Parkes. Chicago: University of Chicago Press.

Brook, Isis (2003) 'Making it Here like There: Place Attachment, Displacement and the Urge to Garden'. *Ethics, Place and Environment* 6: 227–34.

—— and Brady, Emily (2003) 'Topiary: Ethics and Aesthetics'. *Ethics and the Environment* 8: 128–41.

Brown, Jane (2000) *The Pursuit of Paradise: A Social History of Gardens and Gardening*. London: HarperCollins.

Budd, Malcolm (2002) *The Aesthetic Appreciation of Nature: Essays on the Aesthetics of Nature*. Oxford: Clarendon Press.

Burnett, Frances Hodgson (1994) *The Secret Garden*. London: Puffin.

Čapek, Karel (2003) *The Gardener's Year*, trans. G. Newsome. Brinkworth: Claridge.

Carlson, Allen (2000) *Aesthetics and the Environment: The Appreciation of Nature, Art and Architecture*. London: Routledge.

Casey, Edward S. (1993) *Getting Back into Place: Toward a Renewed Understanding of the Place-world*. Bloomington, Ind.: Indiana University Press.

Church, Thomas D. (1995) *Gardens are for People*. Berkeley: University of California Press.

Cooper, David E. (2002) *The Measure of Things: Humanism, Humility and Mystery*. Oxford: Clarendon Press.

—— (2003*a*) 'In Praise of Gardens'. *British Journal of Aesthetics* 43: 101–13.

—— (2003*b*) *Meaning*. Chesham: Acumen.

—— (2005) 'Nature, Aesthetic Engagement, and Reverie'. *Nordic Journal of Aesthetics*.

—— and James, Simon P. (2005) *Buddhism, Virtue and Environment*. Aldershot: Ashgate.

Cottingham, John (2003) *On the Meaning of Life*. London: Routledge.

Crawford, Donald (1983) 'Nature and Art: Some Dialectical Relationships'. *Journal of Aesthetics and Art Criticism* 42: 49–58.

Crowe, Sylvia (1994) *Garden Design*, 3rd edn. Woodbridge: Garden Art Press.

Davies, David (2004) *Art as Performance*. Oxford: Blackwell.

Dewey, John (1980) *Art as Experience*. New York: Perigree.

Dilthey, Wilhelm (1979) *Selected Writings*, trans H. Rickman. Cambridge: Cambridge University Press.

Dōgen Zenji (1996) *Shobogenzo*, ii, trans. G. Nishijama and C. Cross. London: Windbell.

Downing, A. J. (1991) *Landscape Gardening and Rural Architecture*. New York: Dover.

Elliott, Charles (2002) *The Potting Shed Papers: On Gardens, Gardeners and Garden History*. London: Frances Lincoln.

Flaubert, Gustave (1976) *Bouvard and Pécuchet*, trans. A. Krailsheimer. Harmondsworth: Penguin.

Francis, Mark, and Hester, Randolph T. (1990) (eds.), *The Meaning of Gardens: Idea, Place and Action*. Boston: MIT Press.

Friggieri, Joe (2004) *Tales for our Times*. Malta: Progress Press.

Goethe, J. W. von (1971) *Elective Affinities*, trans. R. Hollingdale. Harmondsworth: Penguin.

Goodman, Nelson, and Elgin, Catherine Z. (1988) *Reconceptions in Philosophy and Other Arts and Sciences*. London: Routledge.

Graham, A. C. (2001) *Chuang-Tzu: The Inner Chapters*. Indianapolis: Hackett.

Grice, H. P. (1989) *Studies in the Ways of Meaning*. Cambridge, Mass.: Harvard University Press.

Hadot, Pierre (1995) *Philosophy as a Way of Life*, trans. M. Chase. Oxford: Blackwell.

Harte, Sunniva (1999) *Zen Gardening*. London: Pavilion.

Hegel, G. W. F. (1975) *Aesthetics*, trans. J. Knox, Oxford: Clarendon Press.

Heidegger, Martin (1975) *Poetry, Language, Thought*, trans. A. Hofstadter. New York: Harper & Row.

—— (1980) *Being and Time*, trans. J. Macquarrie and E. Robinson. Oxford: Blackwell.

Hepburn, Ronald W. (2001) *The Reach of the Aesthetic: Collected Essays on Art and Nature*. Aldershot: Ashgate.

Herrigel, Gustie L. (1999) *Zen in the Art of Flower Arrangement*, trans. R. Hull. London: Souvenir.

Hesse, Hermann (2000) *The Glass Bead Game*, trans. R. and C. Winston. London: Vintage.

Holmes, Caroline (2001) *Monet at Giverny*. London: Cassell.

Howard, Elizabeth Jane (1991) *Green Shades: An Anthology of Plants, Gardens, and Gardeners*. London: Macmillan.

Hrdlićka, Z., and Hrdlićka, V. (1989) *The Art of Japanese Gardening*. London: Hamlyn.

Hunt, John Dixon (1998) 'Gardens: Historical Overview'. In M. Kelly (ed.), *Encyclopedia of Aesthetics*, ii, New York: Oxford University Press, 271–4.

—— and Willis, Peter (1988) (eds.), *The Genius of the Place: The English Landscape Garden 1620–1820*. Boston: MIT Press.

Inwood, B., and Gerson, L. (1988) (eds.), *Hellenistic Philosophy: Introductory Readings*. Indianapolis: Hackett.

James, Simon P. (2004) *Zen Buddhism and Environmental Ethics*. Aldershot: Ashgate.

Jekyll, Gertrude (1991) *The Gardener's Essential Gertrude Jekyll*, ed. E. Lawrence. London: Robinson.

Jellicoe, Geoffrey, and Jellicoe, Susan (1995) *The Landscape of Man: Shaping the Environment from Prehistory to the Present Day*, 3rd edn. London: Thames & Hudson.

Ji Cheng (1988) *The Craft of Gardens*, trans. A. Hardie. New Haven: Yale University Press.

Kant, Immanuel (1952) *The Critique of Judgement*, trans. J. Meredith. Oxford: Clarendon Press.

Kass, L. R. (1994) *The Hungry Soul: Eating and the Perfecting of our Nature.* New York: Macmillan.

Keane, Marc P. (1996) *Japanese Landscape Design.* Boston: Tuttle.

Keen, Mary (2001) 'Gardens as Theatre'. In E. Hunningher (ed.), *Gardens of Inspiration*, London: BBC Worldwide, 108–23.

Kemal, Salim, and Gaskell, Ivan (1993) (eds.), *Landscape, Natural Beauty and the Arts.* Cambridge: Cambridge University Press.

Kendall, Richard (1994) (ed.), *Cézanne by Himself.* Boston: Little, Brown & Co.

Kupperman, Joel (1999) *Learning from Asian Philosophy.* New York: Oxford University Press.

Landsberg, Sylvia (n.d.) *The Medieval Garden.* London: British Museum Press.

Leddy, Tom (2000) 'A Conversation between Tom Leddy and Richard Whittaker'. In *Works and Conversations*, iii, <http://www.conversations.org/tom_leddy_2000.htm>.

Long, A., and Sedley, D. (1987) (eds.), *The Hellenistic Philosophers*, i. Cambridge: Cambridge University Press.

Macfarlane, Robert (2004) *Mountains of the Mind: A History of a Fascination.* London: Granta.

McGuiness, B. F. (1990) *Wittgenstein: A Life.* Harmondsworth: Penguin.

McHarg, Ian (1990) 'Nature is More than a Garden'. In Francis and Hester (1990), 34–7.

MacIntyre, Alasdair (1982) *After Virtue: A Study in Moral Theory.* London: Duckworth.

Marcel, Gabriel (2001) *The Mystery of Being*, 2 vols., trans. G. Fraser. South Bend, Ind.: St Augustines's Press.

Matthews, Robin (1990) 'The Trail of the Serpent: A Theological Enquiry'. In Francis and Hester (1990), 46–52.

Merleau-Ponty, Maurice (2002) *Phenomenology of Perception*, trans. C. Smith. London: Routledge.

—— (2004) 'Cézanne's Doubt'. In T. Baldwin (ed.), *Maurice Merleau-Ponty: Basic Writings*, London: Routledge, 272–89.

Miller, Mara (1993) *The Garden as Art.* Albany, NY: SUNY Press.

—— (1998) 'Gardens as Art'. In M. Kelly (ed.), *Encyclopedia of Aesthetics*, ii, New York: Oxford University Press, 274–80.

Mirbeau, Octave (1991) *The Torture Garden*, trans. M. Richardson. Sawtry: Dedalus.

Molesworth, Charles (1995) 'Symbol'. In D. E. Cooper (ed.), *A Companion to Aesthetics*, Oxford: Blackwell, 412–14.

Moore, Charles W., Mitchell, William J., and Turnbull, William (1993) *The Poetics of Gardens*. Boston: MIT Press.

Morris, Ivan (1979) *The World of the Shining Prince: Court Life in Ancient Japan*. Harmondsworth: Penguin.

Murasaki Shikibu (1981) *The Tale of Genji*, trans. E. Seidenstecker. Harmondsworth: Penguin.

Murdoch, Iris (1997) *Existentialists and Mystics: Writings on Philosophy and Literature*. Harmondsworth: Penguin.

Nicolson, Harold (1968) 'Introduction'. In Peter Coats, *Great Gardens of the Western World*, London: Hamlyn, 8–13.

Nietzsche, Friedrich (1968) *Genealogy of Morals*. In *Basic Writings of Nietzsche*, trans. W. Kaufmann. New York: Random House.

Olmsted, Frederick Law (1852) *Walks and Talks of an American Farmer in England*. New York: Putnam.

Osler, Mirabel (2001) 'Symphonic Movements'. In E. Hunningher (ed.), *Gardens of Inspiration*, London: BBC Worldwide, 157–73.

Page, Russell (1995) *The Education of a Gardener*, 3rd edn. Brighton: Harvill.

Perry, Gill (2003) 'The Expanding Field: Ana Mendieta's *Silueta* Series'. In J. Gaiger (ed.), *Frameworks for Modern Art*, New Haven: Yale University Press, 153–206.

Pliny (1963) *The Letters of the Younger Pliny*, ed. B. Radice. Harmondsworth: Penguin.

Pollan, Michael (1996) *Second Nature: A Gardener's Education*. London: Bloomsbury.

Pope, Alexander (1994) *Essay on Man and Other Essays*. New York: Dover.

Richardson, Tim (2005) 'Psychotopia'. In N. Kingsbury and T. Richardson (eds.), *Vista: The Culture and Politics of the Garden*, London: Frances Lincoln. (Typescript version)

Richaud, Frédéric (2000) *Gardener to the King*. trans. B. Bray. London: Harvill.

Robinson, William (1979) *The Wild Garden*. London: Scolar Press.

Robson, David (2002) *Geoffrey Bawa: The Complete Works*. London: Thames & Hudson.

Ross, Stephanie (1998) *What Gardens Mean*. Chicago: University of Chicago Press.

Rousseau, Jean-Jacques (1992) *The Reveries of the Solitary Walker*, trans. C. Butterworth. Indianapolis: Hackett.

Schama, Simon (1995) *Landscape and Memory*. New York: Knopf.

Schimmel, Annemarie (1992) *I Am Wind, You Are Fire: The Life and Work of Rumi*. Boston: Shambhala.

Schopenhauer, Arthur (1969) *The World as Will and Representation*, ii, trans. E. Payne. New York: Dover.

Scruton, Roger (2000) *Perictione in Colophon*. South Bend, Ind.: St Augustine's Press.

Seneca (1917–25) *Ad Lucilium Epistulae Morales*, 3 vols., trans. R. Gummere. London: Heinemann.

Sen Gupta, Kalyan (2005) *The Philosophy of Rabindranath Tagore*. Aldershot: Ashgate.

Slawson, David A. (1991) *Secret Teachings in the Art of Japanese Gardens: Design Principles, Aesthetic Values*. Tokyo: Kodansha.

Smit, Tim (2000) *The Lost Gardens of Heligan*. London: Orion.

Strong, Roy (2000*a*) *Garden Party: Collected Writings 1979–99*. London: Frances Lincoln.

—— (2000*b*) *Gardens through the Ages (1420–1940)*. London: Octopus.

Takei, Jirō, and Keane, Marc P. (2001) (trans.), *Sakuteiki: Visions of the Japanese Garden*. Boston: Tuttle.

Thomas, Keith (1984) *Man and the Natural World: Changing Attitudes in England 1500–1800*. Harmondsworth: Penguin.

Thoreau, Henry David (1886) *Walden*. London: Scott.

Verdi, Richard (1992) *Cézanne*. London: Thames & Hudson.

Wang, Joseph (1998) *The Chinese Garden*. Hong Kong: Oxford University Press.

Wharton, Edith (1988) *Italian Villas and their Gardens*. New York: Da Capo Press.

Wheeler, David (1998) (ed.), *The Penguin Book of Garden Writing*. Harmondsworth: Penguin.

Wittgenstein, Ludwig (1969) *Philosophical Investigations*, trans. G. Anscombe, 3rd edn. London: Macmillan.

Woolf, Virginia (1962) *A Haunted House and Other Short Stories*. London: Hogarth.

Young, Julian (2002) *Heidegger's Later Philosophy*. Cambridge: Cambridge University Press.

Zōen (1991) *Illustration for Designing Mountain, Water and Hillside Field Landscapes*, trans. D. Slawson (see Slawson 1991).

INDEX